SMALL WIND

SMALL WIND

Planning and Building Successful Installations

R. NOLAN CLARK

Amsterdam • Boston • Heidelberg • London
New York • Oxford • Paris • San Diego
San Francisco • Singapore • Sydney • Tokyo

Academic Press is an imprint of Elsevier

Academic Press is an imprint of Elsevier
225 Wyman Street, Waltham, MA 02451, USA
The Boulevard, Langford Lane, Kidlington, OX5 1GB, UK

Notices

Knowledge and best practice in this field are constantly changing. As new research and experience
broaden our understanding, changes in research methods, professional practices, or medical treatment
may become necessary.

Practitioners and researchers must always rely on their own experience and knowledge in evaluating and
using any information, methods, compounds, or experiments described herein. In using such information
or methods they should be mindful of their own safety and the safety of others, including parties for
whom they have a professional responsibility.

To the fullest extent of the law, neither the Publisher nor the authors, contributors, or editors, assume any
liability for any injury and/or damage to persons or property as a matter of products liability, negligence
or otherwise, or from any use or operation of any methods, products, instructions, or ideas contained in
the material herein.

Library of Congress Cataloging-in-Publication Data
Clark, R. N. (Ray Nolan)
   Small wind : planning & building successful installations/R. Nolan Clark.
      p. cm.
   Includes bibliographical references.
   ISBN 978-0-12-385999-0
1. Wind energy conversion systems–United States–Design and construction. 2. Wind power plants–
United States–Design and construction. 3. Wind power–United States. I. Title.
   TJ820.C53 2013
   621.31'2136–dc23

                                        2012023443

British Library Cataloguing-in-Publication Data
A catalogue record for this book is available from the British Library.

ISBN: 978-0-12-385999-0

For information on all Academic Press publications visit
our website at http://store.elsevier.com

DEDICATION

This book is dedicated to all the men and women who contributed their time, expertise, and passion for small wind systems in writing the International Standard "Design Requirements for Small Wind Turbines, IEC 61400-2" and the British and American "Small Wind Turbine Performance and Safety Standards." These standards have established meaningful criteria for engineering design and standardized performance data to allow consumers to make informed purchases. Thank all of you for your hard work and determination in creating these important standards.

CONTENTS

PREFACE

The word *small* is a relative term in that it often means something different to almost everyone. The size of small wind turbines has been changing since the beginning of the use of wind power to grind grain and pump water. Wind turbine sizes in today's small wind category include all but a handful of machines that were constructed prior to 1980. Some notable large wind turbines built before 1980 include the Smith–Putman built in 1941, the John Brown built in 1955, and the Gedser erected in 1957. Several countries in Europe were experimenting with larger wind turbines in the 1950s, but abandoned most development because of the abundance of low–cost petroleum. Many of the first wind turbines installed in wind farms in California, Denmark, and Germany during the mid-1980s had rotor diameters of approximately 15 m, which gave them a rotor area of just under 200 m² (2,200 ft²). Everyone considered them large because they were much larger than the common electric battery–charging turbines sold in the 1930s and 1940s.

The first small wind turbine standard, developed by the International Electrotechnical Commission (IEC) in 1996, defined small wind turbines as those with a rotor area smaller than 40 m² (440 ft²). They had a 7.2 m (24 ft) diameter and a power rating of approximately 13 kW at an 11 m/s wind speed. However, as wind turbines became larger in the early 2000s, the upper limit of small wind was questioned. When the IEC revised the small wind design standard in 2006, the rotor area was increased to 200 m² (2,200 ft²), five times the earlier upper limit. A turbine with an area of 200 m² has a rotor diameter of 16 m (52 ft) and a power rating of 65 kW at a wind speed of 11 m/s. The British Wind Energy Association Small Wind Turbine Performance and Safety Standard also set the upper limit at 200 m². In 2009, when the AWEA Small Wind Turbine Performance and Safety Standard was adopted, the rotor area for small wind machines followed the IEC standard of 200 m². Not too long after that, the association changed their definition of small wind to include all turbines with a rotor capacity of 100 kW or less. A 100 kW wind turbine has a rotor area of 350 m² (3,850 ft²) and a rotor diameter of 21 m (70 ft).

So the upper limit of small wind turbines has grown. As a result, the definition of "small" keeps changing as the wind industry matures and gains more experience with larger machines. The IEC has prepared a draft third revision for the small wind turbine design standard. In the new draft they retain the 200 m^2 upper limit for small wind turbines, which means that, for the next 10 years, we can expect the upper limit to remain the same as is currently.

There are many things to consider when choosing a small wind system either to provide electricity where none is available or to offset the purchase of expensive electricity. The two most important things to consider are the location of the turbine and the type of turbine to purchase. Selecting a proper site requires time to examine the wind speeds available, the permits required, and the land available to meet all safety and operational concerns. Selecting a wind turbine of sufficient size requires knowledge of the anticipated electric load that it will power, as well as knowing that it has been tested and is certified to meet performance and safety standards. It is hoped that the information in this book will help readers make the right choices and guide them in completing a successful installation.

ACKNOWLEDGMENTS

The assistance of Trudy Forsyth and Frank A. Oteri is greatly appreciated in writing the chapters on distributed wind systems and economic considerations. Trudy is retired from the National Renewable Energy Laboratory (NREL)/National Wind Technology Center, where she was a mechanical engineer V and the Distributed Wind Program lead. Frank is a contractor with Wind Powering America/NREL serving as communications specialist.

I thank all the engineers, scientists, manufacturers, policy makers, and research program managers for their candid discussions, criticism, and encouragement over the last 35 years as I led a small team conducting research and testing for the USDA–Agricultural Research Service in Bushland, Texas. I am grateful for all the opportunities I had to conduct research and development work with the engineers at Sandia National Laboratories and NREL and for their support for my research activities with mostly small wind systems.

I am especially thankful that the USDA–Agricultural Research Service developed a cooperative research program with the Alternative Energy Institute at West Texas A&M University at the beginning of the wind energy research program. Dr. Vaughn Nelson's contribution to the research activities was superb, along those of with Ken Starcher and many students.

I am deeply indebted to the engineering staff that work with me at the USDA–Agricultural Research Service in Bushland, especially Fred Vosper, Fadi Kamand, and Ron Davis, who were there in the early years as we began the wind energy research program. Shitao Ling and Eric Eggleston helped in the development of small wind turbine controllers for stand-alone water pumping and hybrid systems. In later years, Brian Vick, Adam Holman, and Byron Neal helped with turbine blade testing and the beginning of certification testing.

I express my gratitude to my wife Ann for her patience with me while I wrote this book, along with her willingness to travel the side roads of America looking at wind turbines, sometimes spending hours waiting until I finished visiting with manufacturers or turbine owners or operators.

Harvesting the Wind

The energy in the wind was first used to power small boats. As man learned to control these crafts, he built larger boats with larger sails. In time, sails were placed on rotating booms to turn shafts and thus operate grinding stones and pumps. Archeological records have shown that early forms of windmills were used by the Babylonians, Chinese, and Egyptians. Written references to windmills are found in early manuscripts, with a Persian vertical-axis windmill described in some detail in the 7th century A.D. European documents after the 13th century refer to various windmill designs. European mills used rotors or sails constructed primarily of wood, reeds, and canvas prior to 1900. Many different concepts and designs were used, including springs and shutters to increase or decrease the sail area. Peak power from some of the larger mills reached 30 kW [1]. Figure I.1 shows a restored 18th-century Dutch windmill used to crush and grind seeds to produce vegetable oils.

Windmill development in the United States duplicated European designs, but American machines did not provide the flexibility needed to withstand the fickle weather of the Midwest. In 1857, Daniel Holladay

Figure I.1 Restored 18th-century Dutch windmill used for grinding grains. Others were used for pumping water.

Small Wind
ISBN 978-0-12-385999-0

1

began making wind machines that were self-regulating using paddle-shaped blades that pivoted, or feathered, as wind speed increased. The Eclipse windmill was introduced a few years later and was the first to use a solid wheel assembly and a side vane to turn the rotor out of the wind as velocity increased. Both of these machines used a reciprocating pump driven by a crank or offset cam. Such wind machines worked sufficiently to lift water from deep underground sources in the arid parts of the United States. Enclosed gears, metal wheels, and towers improved until the systems operated well in light winds and ran smoothly. At the beginning of the 20th century, it is estimated that 200 U.S. companies were offering wind-mills for powering saws and shelling corn as well as pumping water. Figure I.2 shows some of these windmills, which have been restored and placed in a windmill museum in Texas.

Wind machines that generated electricity for charging batteries were being manufactured in the 1930s. They were much different from earlier multiple-blade water pumpers in that their two or three blades rotated at much higher speeds. Electrical output was normally 12, 24, 32, or 48 volts DC, and they incorporated batteries for energy storage. Most electric systems were capable of generating and storing enough power to operate two to three lights and a radio. The Jacobs Wind Electric Company reported selling tens of thousands of these units between 1931 and 1957 [2]. Most of the electric systems were discarded when the U.S. Rural Electrification Administration (REA) installed electric power lines in many farming and

Figure I.2 Restored American windmills manufactured from 1875 until 1940 and used for water pumping. This basic design is still being manufactured.

ranching regions. However, water–pumping windmills never disappeared from the western United States because they were needed to provide water for livestock in remote areas.

During this time, several countries in Europe were experimenting with larger wind turbines that would generate electricity for connection to the electric grid. Almost all of these efforts were abandoned sometime in the 1960s because of the abundance of low–cost petroleum for fueling steam–electric plants. However, the 1973 oil embargo resulted in a desire to develop alternative energy sources, especially to generate electricity. Governments in Europe and the United States started development programs using new aerodynamic theories and manufacturing techniques to produce new wind machines. At the same time, wind enthusiasts began to restore old wind electric plants to be used for electric power generation. By 1984, some 8,000 new units were installed and had a combined generation capacity of 300 MW [2]. They were manufactured by over 60 companies. Many were installed in Denmark, Germany, and California. The European machines were typically three-bladed, up–wind machines with a rotor diameter of 15 m (50 ft). They had a power rating of approximately 65 kW. U.S. manufactures produced either two–bladed or three-bladed down–wind machines with rotor diameters ranging from 10 m (33 ft) to 17 m (56 ft). By 1990, rotor diameters had more than doubled in size and power ratings had quadrupled.

The focus of most government–sponsored research and development was on wholesale electricity to utility grids. Little public money was used to advance designs of small machines used for homeowners and businesses. In the United States, the Department of Energy and the Department of Agriculture jointly sponsored some development activities related to applications in rural and remote areas, but little effort was put into new wind machine design. European countries focused on developing systems for export to developing countries with no electric grid and concentrated on battery charging and water pumping.

Thus most new development has been supported and carried out by individual wind turbine manufacturers rather than government programs. Beginning in the mid–1990s, The Department of Energy funded some research and development through the National Renewable Energy Laboratory (NREL), but individual manufacturers still did most of the work. Table I.1 lists early manufacturers of small wind machines, some of which have survived the ever–changing marketing and regulatory environment. Many other manufacturers closed after producing only a few machines,

Table 1.1 Examples of small wind turbines manufactured during the 1980s

Manufacturer	Model	Rotor diameter (m)	Rated power (kW)	Generator type
Aerolite	1200	7.0	12.5	Induction 120/240 AC
Aerowatt (France)	24 (other models up to 900 watts)	1.2	30 watts	12 volts DC
Bergey Windpower	1000	2.5	1.0	Permanent magnet alternator DC or AC
	Excel	7.0	10.0	200 volts DC
Dakota Wind & Sun		4.3	4.0	Induction 120 AC
Enertech	1500	4.0	1.5	Synchronous AC/DC
Jacobs Wind Energy Systems	10	7.1	10.0	Synchronous AC
	15	7.4	15.0	Synchronous AC
	17.5	8.0	17.5	12 volts DC
Marlec Engineering Company (UK)	Rutland 500	0.51	50 watts	
North Wind Power (Northern Power Systems)	HR2	5.0	2.2	Synchronous alternator with wound stator
Southwest (Wind Baron)	Windseeker	1.52	350 W	Permanent magnet alternator DC
Whirlwind	2012	3.4	2.0	120 volts DC
World Power Technologies	Whisper 600	2.1	600 Watts	Permanent magnet alternator DC or AC
	Whisper 1000	2.8	1.0	
	Whisper 3000	4.5	3.0	

mostly because of insufficient capital to finance the several iterations needed to create a successful product. Still others have been subject to sale or merger.

At least three major design changes have influenced the performance and reliability of small wind machines over the last 20 years. The switch from DC generators and small induction motors to permanent magnet alternators has significantly improved performance and extended lifetime operation (Figure I.3). Alternators can be built in small compact units that are lighter and can be easily fitted to a small tower. A second design change is in rotor blade materials and construction. The use of injection-molded blades provides a much more efficient airfoil and one with better strength properties to withstand extreme wind speeds and higher operating speeds (Figure I.4). Lastly, the inverters and power electronics available for small wind machines provide almost seamless integration with the electricity grid (Figure I.5). They provide systems that have almost 100% availability and perform real-time monitoring of the wind machine.

Although small wind turbines do not make up a large percentage of new generating capacity, they do account for more than a doubling of the number of medium and large turbines installed. In 2011 almost 7,500 small wind turbines were installed in the United States, with U. S. manufacturers exporting slightly more than 3,000 units [3]. Of the number of small wind turbines installed, only 4% were larger than 10 kW. A little more than 5,000

Figure I.3 The development of permanent magnet alternators has provided improved performance and reduced maintenance.

Figure I.4 The development of injected molded rotor blades has provided more efficient airfoils and improved structural properties.

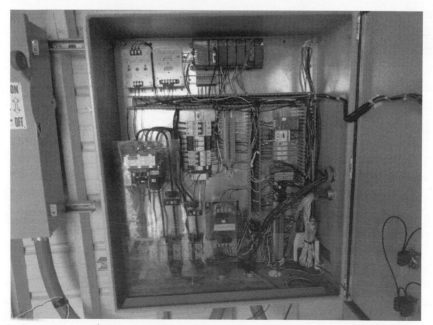

Figure I.5 Power electronics and smaller computer chips have provided the components to develop controllers that maximize the performance of wind turbines.

were units less than 1 kW in size and about 2,000 were in the 1 to 10 kW range. In terms of small wind systems, then, they are largely less than 10 kW and have a rotor area smaller than 40 m² (440 ft²).

The number of grid–connected small wind turbines has been increasing since 2008 in the United States and reached almost 3,000 units in 2011 [3]. At the same time, the number of off-grid systems dropped from 7,600 in 2008 to 4,500 during 2011. Part of the change from off-grid to on-grid turbines is the introduction of several new wind turbines in the 2 to 5 kW range that are designed for the on-grid market. Also, the economic downturn beginning in 2009 has reduced the total number of small wind turbine purchases.

The world market for small wind turbines remains strong with a growth rate of 26 percent. It is estimated that there are over 300 small wind turbine manufacturers in the world, with Canada, China, Germany, the United Kingdom, and the United States producing about half of them [4]. The average size of small wind turbines installed in China was 0.37 kW in 2010 while the average size was 1.24 kW in the United States and 2.0 kW in the United Kingdom. These differences are largely due to the number of turbines being installed as grid-connected versus off-grid units.

WIND ENERGY PRINCIPLES

The main principle of wind energy is the harvesting of kinetic energy in the wind by a rotor or wheel and transferring it to a rotating shaft. The energy in the shaft is then used to directly pump water, drive an electric generator, or produce heat. Therefore, wind energy is normally converted to mechanical, electrical, or heat energy. Power is a concept frequently used in describing the performance of wind machines; it is a measure of the energy extracted during a specific period of time. The power in the wind is dependent on the volume of air that passes through the rotor area perpendicular to the wind per unit of time. The theoretical power in a wind stream is determined by

$$P = \tfrac{1}{2}\,\rho A V^3$$

where P is the power in watts, ρ is the air density in kilograms per cubic meter, A is the cross-sectional area in square meters, and V is the wind speed in meters per second. (It is best to calculate the power with metric units rather than try to balance all the conversion factors required for English units.) The

actual energy extracted from the wind stream is always less than this theoretical amount because harvesting 100 percent of the energy would require
that no wind occur behind the rotor. Experiments have proven that the
maximum energy extractable from the wind is 59.3% of the power available.
In actual practice, a wind rotor captures significantly less than the maximum
because of friction in the bearings, generator, and other rotating objects.

Efficiency, or power coefficient (c_p), is determined by

$$c_p = \text{Power delivered}/\tfrac{1}{2}\,\rho A V^3$$

The power coefficient is dependent on the design of the system's airfoils as
well as the friction of all of its rotating components. The effects of turbine
design on the power coefficient are discussed in Chapter 4, Wind Turbine
Components and Descriptions. This efficiency should not be confused with
another term that measures the performance of a wind system over an
extended period of time, the capacity factor. The capacity factor indicates
performance in relation to a predicted maximum energy production. It
includes not only the efficiency of the turbine but also the deviation of the
wind resource from the normal. The capacity factor varies from month to
month depending on the variance of the wind resource, whereas the efficiency or power coefficient does not vary from month to month unless there
is significant wear in some component. The capacity factor is best used to
describe the performance of a group of wind machines such as in a wind farm
rather the performance of a single machine in a dedicated application.

Historically, we have grouped wind machines more by application than
by design or size, but over the last 20 to 25 years that has changed as wind
machines have gotten significantly larger and more and more machines are
being connected to the electric grid. Generally, wind machines today are
grouped as small or large, with a new mid–size grouping entering the
discussion. However, since almost 100 percent of machines in the mid–size
and large categories are grid–connected to produce mostly wholesale electricity, categorization by application is quickly disappearing. For machines
that are classified as small, there are still divisions by application because
battery charging, water pumping, and electrical grid connections require
machines that are different in many ways. Therefore, it is important to keep
the application firmly in mind when discussing and considering small wind
machines for purchase and installation.

Many manufacturers and installers prefer to discuss small wind machines
in terms of size because they are looking to match the wind system to
a particular load or application. They find it much easier to offer

a prospective customer the outcome of a machine based on its power level rather than its rotor area. However, safety and performance standards have selected rotor area as a means of describing the different wind systems available. Most current safety and performance standards use 200 m^2 (2,200 ft^2) as the maximum size of small wind machines. A wind turbine with a rotor area of this size has a rotor diameter of approximately 16 m (52 ft) and a power rating of 65 kW at a wind speed of 11 m/s (24.6 mph). Although small wind machines do not make up a significant percentage of new generating capacity, they do account for more than double the number of medium and large turbines installed.

Small wind machines are usually encountered in one of three installation types. Many are installed in stand-alone applications where there is no electric utility power available, and may be for battery charging, water pumping, or remote hybrid systems. More and more small machines are being installed on the consumer or load side of the electric meter. They reduce the amount of electricity purchased from the utility at retail prices. The introduction of several machines in the 2 to 5 kW range has increased the potential of the on-grid market. A few small machines have been installed to provide power to the wholesale electric grid market. These are often located in areas where the erection of larger machines is prohibited by either regulations or availability of adequate erection equipment. Installing a machine on a remote island is an example.

REFERENCES

[1] Golding EW. The Generation of Electricity by Wind Power. London: E. & F. N. Spon; 1955.
[2] Nelson VA. History of the SWECS industry in the U. S. Alternative Sources of Energy, No. 66, 1984; p. 20.
[3] American Wind Energy Association. 2011 U.S. Small Wind Turbine Market Report; June 2012.
[4] World Wind Energy Association. 2012 Small Wind World Report Summary. WWindEA.org; March 2012.

CHAPTER ONE

Site Evaluation
Examining the Proposed Site to Ensure That It Has Adequate Wind and Space

Selecting a proper location for your wind energy system is probably more important than deciding which wind system to purchase because a well-designed wind machine doesn't operate at its performance capabilities if it has poor wind conditions. Many wind energy systems have not been successful economically because of poor site selection. The site selection process not only involves the physical location for best wind and safety concerns, but involves placing the wind turbine at the proper height for reduced turbulence and good wind. Site selection begins with determining the wind speed at the desired location and then involves considering safety issues and noise considerations.

Determining the wind speed at a proposed wind turbine site is critical to estimating the economic potential of the wind turbine. The preferred way is to measure the wind speed with accurate wind-recording equipment. However, this is often cost-prohibitive for small wind systems because of the relative cost of the measuring equipment to the actual cost of the wind-generating system. Typically, you do not want to spend more than about 5% of the cost of the wind system in determining the wind speed at a proposed site. A basic wind-recording system, which includes tower, anemometers, and data logger, is about $7,500.00 [1]; therefore, the estimated cost of the wind system would need to be above $150,000 before you wanted to spend the money to do actual wind measurements. However, if the proposed system is for a public project such as a school or government building, it might need to have a wind measurement program to provide a more accurate estimate of the performance.

There are several other ways of estimating the wind at any potential wind turbine location in addition to making actual measurements. The U.S. Department of Energy (DOE) published their first wind energy atlas in 1986. The staff at the National Renewable Energy Laboratory has continued to

Small Wind
ISBN 978-0-12-385999-0

upgrade these wind data into wind energy maps and has made them available at the DOE-sponsored Wind Powering America web site [2]. The web site address is www.windpoweringamerica.gov/windmaps. The site contains wind maps for utility-scale wind systems with maps at 80 m height for land based systems and 90 m height for offshore systems. The maps for community-scale wind systems are at a height of 50 m. These maps should be used for mid-size turbines as well. The last set of maps are for residential-scale systems or small wind systems and are at a height of 30 m. Figure 1.1 is the U.S. map showing the wind speeds at 30m. You can access individual state maps by clicking on the state that you are interested in and the state map will appear. There is a download icon so that the map can be downloaded to your computer. Figure 1.2 is a map for the state of Iowa. The state maps show more detail and it is easier to locate the specific area or areas of interest. Once the maps are downloaded, you can zoom in to have a closer look at areas of interest as well. These maps are continually being updated by the staff at the National Renewable Energy Laboratory, so these should provide a good estimate of the wind speed at any

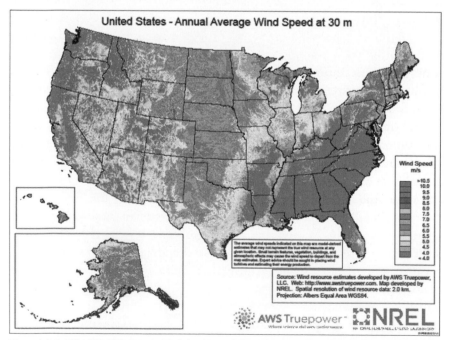

Figure 1.1 Wind resources map of the United States for a height of 30 meters *(National Renewable Energy Laboratory, 2012)*

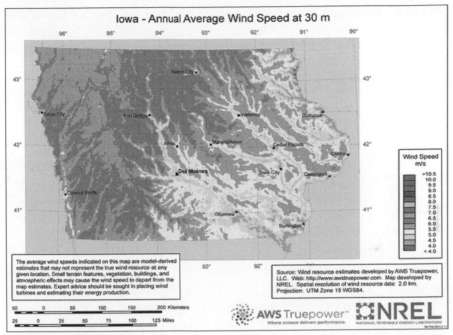

Figure 1.2 Wind resources map of Iowa for a height of 30 meters *(National Renewable Energy Laboratory, 2012).*

location. Wind maps are available for many countries throughout the world. The National Renewable Energy Laboratory web site has wind maps available for about 30 countries. The Riso National Laboratory in Denmark also maintains wind maps for almost all of Europe and several different countries. These wind maps give either average wind speeds or wind power densities for many regions. Because worldwide maps are available at various heights, you need to be sure to note the height of the wind resource information obtained. Wind speed varies with height and will need to be corrected to the proposed height of your wind system.

You may also obtain an average wind speed from some nearby source such as a weather station, airport, or any number of agencies or businesses that collect local weather. If you obtain local wind data be sure to ask at what height data were collected because you will probably need to correct it for height. The change in wind speed with height is called wind shear. It has been measured in all types of weather conditions and can be either increasing as height is increased or it can, under certain weather conditions, decrease with height. Wind shear can be calculated by using a power law function

$$v = v_o\{H/H_o\}^{\alpha},\qquad(1.1)$$

where v is the new wind speed at height H, v_o is the measured wind speed at height H_o, and α is the wind shear exponent. Under what is called stable conditions, where air temperature decreases with height, the wind shear exponent will normally be 1/7 or 0.14. The wind shear exponent will vary depending on atmospheric conditions and on the terrain or roughness of the ground. One thing to consider in determining wind shear near homes and businesses is that as landscaping grows over the years, the wind shear will usually increase and the wind speed will be reduced at the hub height of the wind turbine. Many measurements of wind shear made between 10 and 40 meters show that the power function is nearer 0.20 than the usually assumed 1/7 or 0.14 [3]. Although the effect of wind shear on wind speed seems slight, one must remember that since wind power is a function of the wind speed cubed, small changes in wind speed can affect wind power greatly. Figure 1.3 shows changes in wind speed as height increases beginning with a wind speed of 10 m/s at a height of 10 meters. The wind speed at 50 meters is 12.6 m/s, which produces approximately twice the power as a wind speed of 10 m/s, which occurs at a 10-meter height. Care must be taken in using wind data collected at many weather stations, normally at a 10-meter height, and using the wind shear component to estimate wind speeds at 30- and 40-meter heights because conditions vary so much that the wind shear exponent may not be that representative of the location. This is why it is

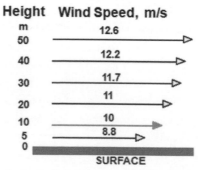

Figure 1.3 Changes in wind speed at various heights above the same surface showing effects of wind shear on wind speed.

always best to measure the wind speed at the site and at the height of the proposed wind turbine.

CALCULATING WIND SPEED DISTRIBUTION FROM AVERAGE WIND SPEED

If one knows the average wind speed for a location and has corrected it to the desired height, then a wind speed distribution can be calculated using either Weibull or Rayleigh distribution. Rayleigh distribution is a special case of Weibull distribution and is used for most land-based wind turbine sites. The International Small Wind Turbine Standard IEC 61400-2 [4] recommends use of the Rayleigh distribution which is given as

$$F(v) = \Delta v \, \pi/2 \, v/v_a{}^2 \exp\{-\pi/4[v/v_a]^2\}, \qquad (1.2)$$

where $f(v)$ is the frequency of occurrence associated with each wind speed, v; which is at the center of Δv; Δv is the class or wind speed bin; and v_a is the average or mean wind speed. The wind speed histogram can be calculated for one year by 8,760 times $F(v)$. The wind speed histogram for a month can be calculated using $F(v)$ for the month times the hours in the selected month.

Figure 1.4 shows an annual wind speed histogram calculated from the Rayleigh distribution formula compared to a measured annual histogram

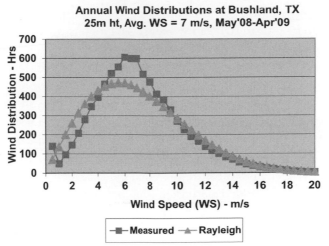

Figure 1.4 Actual measured wind speed distribution and calculated Rayleigh wind speed distribution for Bushland, Texas. Data collected at a 25-meter height in 2008 and 2009.

at Bushland, Texas. The average wind speed for this height and years was 7.0 m/s. Note that the actual histogram has some humps and valleys compared to the smooth curves of the Rayleigh histogram. This will always be the case when comparing a calculated histogram to a measured histogram. Also note that the peak is slightly different and not as sharp. Generally, Rayleigh distribution is a good approximation of the actual histogram. Before calculating a wind speed histogram using Rayleigh distribution, be sure to correct the average wind speed for height. It is much easier to correct average wind speed than trying to adjust wind speed bins and hours.

MEASURING WIND SPEED AT POTENTIAL WIND TURBINE SITE

The most desired method to determine wind speed at a potential wind turbine site is to measure the wind speed at the proposed site and height. A typical wind–measuring system includes the following basic equipment: anemometer, wind vane, data logger, tower, and associated wiring. If possible, a temperature sensor and barometric pressure sensor add important data to allow for air density calculations. This is especially important if the elevation is greater than about 300 meters. Both cup- and propeller-type anemometers are used for wind resource measurements. Cup-type anemometers (Figure 1.5) normally employ a pulse–counting type of signal that can be transmitted easily over several meters with very little loss of signal. This makes the cup anemometer more useful for wind power measurements than the propeller type because direct current (DC) generator outputs cannot be transmitted over very long distances before the signal is diminished. Sonic-type anemometers have been used some in recent years, but are plagued by suspended particles in the air stream. Dust, snow, and other objects transported in the wind give unusually high wind speeds. If the wind is to be measured in areas that experience icing, then a heated anemometer is needed to measure the wind speed accurately at all times. It is usually not worth the added expense if icing only occurs for one or two days per year.

Wind direction is not used to determine potential wind generation, but is important data to have to evaluate seasonal effects on the wind power generated. Many locations have a predominate wind direction in winter and another predominate direction in summer or spring. The more you know

Figure 1.5 Cup anemometer used to measure wind speed.

about wind direction and periods of good or poor winds, the better you can select a site that will maximize your energy production. Wind direction also helps determine or predict the influence of nearby buildings or other potential obstructions.

Another main component of wind-measuring equipment is a data logger. The logger needs to be able to sample the number of channels fast enough to obtain a sufficient number of readings between each averaging time. Typically, wind data need to be sampled at speeds of one sample per second, or faster. Sampling at one sample per second and calculating an average every minute yields a recorded reading that is a composite of 60 individual samples. This provides much better accuracy than recording individual samples every minute or even every five minutes. Development of a wind speed histogram is the desired outcome of collecting wind speed data at the potential wind turbine site.

ESTIMATING ANNUAL ENERGY PRODUCTION

With a wind speed histogram, fairly accurate estimates of annual energy production can be made for any wind turbine using the power curve for that wind turbine. If there are large seasonal variations in wind speeds, then

Table 1.1 Calculated wind energy production using wind speed histogram and wind turbine power curve for 5 kW wind turbine

Wind speed (mph)	Power (watts)	Bin hours (hours)	Energy (kWh)
1—8	0	3,181	0
9	0	731	0
10	50	660	33
11	250	592	148
12	500	493	247
13	850	429	365
14	1,100	348	383
15	1,300	277	360
16	1,600	225	360
17	1,800	197	355
18	2,000	159	318
19	2,400	126	302
20	2,900	88	255
21	3,300	63	208
22	3,700	57	211
23	3,850	45	173
24	4,000	31	124
25	4,200	20	84
26	4,400	13	57
27	4,800	8	38
28	4,850	7	34
29	4,900	4	20
30	4,950	1	5
31	5,000	1	5
		7,756	4,085

monthly or seasonal histograms may need to be determined. Table 1.1 shows an annual energy production calculation using a wind speed histogram and wind turbine power curve. The energy produced at each wind speed is calculated and then summed to obtain the annual energy production. Again, remember that the energy production for any time period can be determined using this method.

Another method to estimate annual energy production is to use the wind power density method. Almost all of the wind resource maps published prior to 2012 wind power densities divided into wind classes. The wind power density for these maps was determined by using a Weibull wind speed distribution with a "k factor" of 2.0. When a k factor of 2.0 is used in a Weibull distribution, it becomes exactly like a Rayleigh distribution.

Table 1.2 Wind power density per unit area of wind turbine rotor for various wind speeds (based on Weibull k value of 2.0)

Wind Speed (m/s)	Wind Speed (mph)	Power density (W/m²)
4.4	9.8	100
5.6	12.5	200
6.4	14.3	300
7.0	15.7	400
7.5	16.8	500
8.0	17.9	600
8.4	18.8	700
8.8	19.7	800
9.1	20.1	900
9.4	21.0	1,000

Table 1.2 contains wind power densities for various wind speeds and can be used to estimate the wind power density when the average wind speed is known. The annual energy production can be calculated for any wind turbine if the wind turbine rotor area and efficiency is known. The annual energy production as a function of rotor area and wind power density and is calculated by

$$AEP = CF \times area \times WM \times 8,760/1,000, \tag{1.3}$$

where AEP is the annual energy production in kWh/year, CF is the capacity factor or efficiency factor of the wind turbine, area is the area of the wind turbine rotor in m², WM is the wind power density in W/m², and 8,760 is hours per year. The 1,000 converts watts into kilowatts. As an example, a 5 kW wind turbine with 2.8 m blades produces about 13,000 kWh per year when the wind power density was 250 W/m² and the efficiency factor was 25%.

Using a wind resource map is a quick way to estimate the size of wind turbine needed for a particular location or load and to evaluate a particular location as a potential for using wind energy systems. The wind resource at a particular location can be enhanced by using a taller tower because wind speeds increase with height.

SELECTING THE LOCATION FOR THE WIND TURBINE

Now that you have determined that you have sufficient wind to produce the power needed for your wind turbine, it is time to consider the location of this wind turbine. The majority of small wind turbines are located near the load that they will be powering, which means there will usually be buildings close by. Buildings and trees usually affect wind flow patterns for small wind turbines more than hills or valleys. The terrain is important in wind flow patterns, but these patterns are disrupted by local objects that become more of an influence on small wind turbines. Therefore, it is important to consider the effects of buildings on the wind flow patterns near them. The shape of the building will have an impact on the downwind flow pattern. The disturbed wind flow pattern around a building looks like a horseshoe with the effect extending downwind for 15 to 20 times the height of the building. Trees, any other structures, or natural barriers will have a similar effect as buildings to the wind flow pattern, causing wind eddies and turbulence. Figure 1.6 shows the areas before, above, and after buildings and trees of disturbed wind flow. The area before a structure is rather small compared to the area after the structure. The distance in front or upwind of the structure is approximately two times the height of the object. The distance after or downwind is 20 times the height. The height above the object is two times the height measured from the ground. To avoid eddies and turbulence before and after an object, it is easier to place the wind turbine on a tower tall enough to be above these areas of disturbed wind. Remember that you must consider the various wind directions when evaluating the effects of objects on your wind flow. It is not uncommon for a location to have two or more prevailing wind directions. Coastal locations may have a prevailing daytime wind direction and a

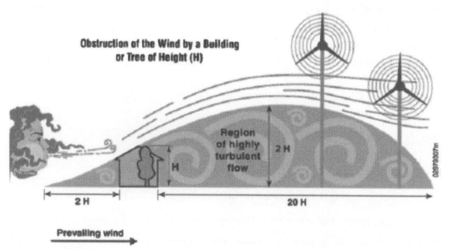

Figure 1.6 Wind flow pattern around and over buildings and trees.

different nighttime prevailing direction. Much of the central United States has a southerly summer prevailing wind direction and a northerly winter time prevailing direction. This is why it is important to measure the wind direction when collecting wind data for a prospective wind turbine location. Another common mistake made in selecting a site for a wind turbine is to not consider any growth of trees or other types of vegetation. If trees are older and mature, they may not grow too much during the next 20 years; however, trees that are younger could easily double in size in 20 years. Increasing a tower from 20 to 30 meters often doubles the energy produced by the turbine. Also, increasing the tower height is often less expensive than moving the turbine further away from the load.

Figure 1.7 contains some estimates of the effects of a building on wind speed, turbulence, and wind power at various distances downwind from the object [6]. The wind speed is decreased by 17% and wind power by 43% at a distance of 5 heights, while the turbulence is increased by 20%. At a distance of 10 heights, the wind speed is decreased by 6%, but power is decreased by 17%. This is still a significant reduction in power at this distance downwind of an object. Even at a distance of 15 heights, the wind power is reduced by 9%. These losses are created just by not locating the wind turbine in an optimum location.

Another consideration when selecting a location is distance from the load. Electrical wiring will need to be run from the wind turbine tower base to the load. It is recommended that all wiring near the wind turbine be buried so that access to the site is not limited due to overhead electrical lines. The area near the turbine base needs to be free of poles, trees, or other

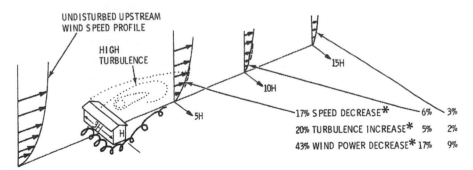

*APPROXIMATE MAXIMUM VALUES DEPEND UPON BUILDING SHAPE, TERRAIN, OTHER NEARBY OBSTACLES

Figure 1.7 Effects of buildings and trees on wind speed, estimated wind power, and turbulence at different distances downwind.

objects that will hinder assembly and erection of the tower and turbine. Plan your electrical wiring route. What objects will have to be crossed? Are you crossing driveways, sidewalks, other utility lines, and so on? Try your best to avoid crossing large objects such as driveways because of the cost and difficulty of maintaining the buried electrical lines. Determine the distance because voltage is lost when electricity is transferred through wires. Larger wires transport electricity with less loss, but the cost can be almost prohibitive if distances are long and voltage is low. Turbines that produce DC voltage are most susceptible to high voltage losses. It is best to use the highest voltage possible when transferring power from the wind turbine to the load or meter. All electrical wiring needs to be installed to meet the local electric code or national code. In the United States, the National Electric Code should be followed to determine the correct wire sizes and safety switches to use. Make sure the wiring system is grounded properly for protection from lightning strikes and stray voltages. All electrical wiring should be installed by trained electrical technicians.

Assuming that the small wind turbine will be a machine that produces electricity used by a nearby load, the electrical output would be either DC voltage passed through an inverter or AC electricity that has the same voltage as the load. In many cases the electricity would be single-phase 120 or 240 volts. However, if the wind machine is larger than 10 to 15 kW, the electricity will probably be three-phase 240 or 480 volts. All of these are considered consumer voltages as opposed to distribution voltages of 13,200 volts or higher.

At this point in the site selection and evaluation process, you need to be talking to the local electrical supplier. The three most common types of electrical suppliers are municipal-owned electric companies, investor-owned electric companies, or electrical cooperatives. It is important to talk with your electric company to obtain their rules and procedures for connecting a wind turbine to their electrical system. The guidelines for electrical connection are slightly different for the three common types of electrical companies, and the application of those guidelines varies from utility to utility. Do not assume that utility A in your town will allow a connection like utility B in the neighboring town. Get all the information about the physical connection and the contract information, which will include initial connection fees, monthly demand or user fees, and finally the cost of purchased energy and the credit for any surplus energy supplied to the electrical grid. You will need all the electrical contract costs to calculate the potential economic feasibility of installing a wind turbine.

INSTITUTIONAL SITING

Not too far into the site evaluation process, one needs to start examining the institutional rules and regulations that might impact the erection of a wind turbine on a selected piece of property. If the site is in a rural area it may not have any restrictions about how the land is used except for some restrictions enforced by the Federal Aviation Agency (FAA). The FAA has a ruling that all towers or structures taller than 61 meters (200 feet) be lighted with approved lights. In most cases this will not be an issue with small wind turbines because most of them will be less than 45 meters (150 feet) tall. However, some states or counties do have land use ordinances that apply to all land within the state or county. Land designated for agricultural use oftentimes has no restrictions on what type of structures can be built; even though there is a land use policy, it does not impact the possibility of erecting a wind turbine. If you plan on locating the turbine on property within a city, town, or village or near one of these urban centers, you will probably find that most have zoning and building codes. It is time to study and understand these rules and regulations thoroughly.

It is important to remember that persons working in the code enforcement offices did not make the rules in the majority of cases. These office workers and field inspectors were simply hired to provide enforcement for the local authority. The actual rules were developed and approved by a board of elected officials who you may never see or talk to so please treat the people with respect and understand that they can be a great help to you in getting approval for your wind turbine. They can become an ally to help you get a variance from the rule for this single project. These officials can help you learn about all the permits that may be required. Building permits, inspections, and approvals are all part of the process to protect the general public and to ensure a level of safety in the neighborhood. For small wind turbines, these local inspectors have the final say about where and how your wind turbine is erected. You must work closely with them and, in many cases, help educate them about the workings and advantages of using wind power. Don't be surprised if you hear "We've never done that before."

During this planning process, it is a good time to start visiting with neighbors and to start getting a feel of their attitude about wind energy and how they would feel if you put up a wind system. It is important to know which neighbors might be ones in favor of a wind turbine in the neighborhood and ones who would be opposed. Try to learn what their objections might be and start to gather information about the truths and myths of that component of the wind system. A good example about wind turbine myths is that they kill many birds. Granted birds are killed by wind turbines, but the number is low compared to those killed by cats and dogs or cars or by flying into buildings [7]. Gather some information about any concerns your neighbors might have so that you can answer their questions or help them understand the nature of wind turbines and their operation. Invite them to go look at a working machine if one is nearby. Don't surprise your neighbors by unloading a machine and starting erection without talking with them.

SETBACK

Another important siting issue for small wind turbines is setback or the distance from the tower base to the property line or to inhabited buildings. Many potential issues are alleviated by using a proper setback distance for the wind turbine. Safety and noise are the two largest issues that can be overcome by using a minimum setback of at least the height of the tower and the length of a single blade. Seldom has the author seen complete towers fall and stay intact and reach out the total length of the tower. More often it is a top section that will fold and fall close to the tower base, then pulling over the remaining tower. Guyed towers are more prone to total failure than freestanding towers. Difficulties with guy cables are often the cause of failure in guyed towers because cable fasteners work loose with time or guy cables are impacted by mowers, vehicles, and so on, causing the cables to break. In a few cases, blades have been ejected from the turbines due to failure of a blade. When blades fail, they usually crack and break from one side, leaving them dangling to strike the tower before falling to the ground. Once a blade hits the tower, it usually falls almost straight down, leaving all debris close by the tower or well within the radius of the tower height.

Another potential safety issue requiring setback is ice flying off the rotor blades. In climates where ice accumulates on objects during wet cold weather, wind turbine blades will ice over unless heated. The rotor blades will ice over when temperatures are slightly below freezing and the atmosphere is saturated with fog or light drizzle. Typically blades do not ice up from precipitation that is already frozen, such as snow or sleet. If the blades are rotating when the temperature goes up, ice will be thrown from the blades. Most of the ice will fall close to the turbine base, but some could be thrown up to a distance equal to the tower height.

All wind turbines make some noise because of the rotation of the blades and the generator. Small wind turbines tend to make more noise than larger machines because the blades turn faster. Because the optimum design for an airfoil is to turn at a tip speed ratio of six to eight times the wind speed, all wind turbines are designed to turn at a rotational speed that will maximize the efficiency of the airfoil. A wind turbine with a diameter of 15 meters (50 feet) will nominally turn about 60 rpm, whereas a small 3-meter (10-feet) rotor will turn about 175 rpm. The generator is another source of noise in wind turbines but usually can be reduced by adding insulation to the generator housing. Another source of noise is the gear box. The rotation of gears creates a steady grinding noise that is fairly consistent and is usually not too much of a nuisance. Because noise is also dissipated with distance, the best way to lower the noise level at your property line is to move the turbine away from the line. Because of changing wind conditions, there will be times that any wind turbine will produce noise levels that will be audible at two to three times the recommended setback distance. These events should be rare and should not be considered a nuisance noise level.

SITING ISSUES FOR CONSTRUCTION

One is always looking at potential sites to determine the proximity to the load and the effects of nearby structures and objects on the wind flow. Once a site is chosen that is close to the load and has good wind flow, you next need to consider the site as a construction site. How much room do you have to work with? First you need access and room to bring in the trucks that will be delivering the turbine and its components. A land owner purchased a small wind turbine (2 kW) with tower and was waiting delivery. The turbine was shipped with the tower already assembled in 9 meter (30 feet) sections. The truck was a standard 18 wheeler but could not turn from the roadway into the owner's narrow driveway. Therefore, the driver simply rolled the tower off the truck into the ditch in front of his house, leaving the new turbine owner with the problem of picking up the tower sections and moving them to the back of his property to the erection site. If a crane will be required for erection, are the roads and bridges capable of supporting the weight of a crane coming to the site? Make sure you

(1) have enough free room at the turbine site to allow for the storage of turbine parts and (2) have room to construct the foundation and assemble the tower and nacelle. Many small turbines will use a tilt-up procedure to erect the tower regardless of whether it is freestanding or guyed. The site has to have room to accommodate construction and assembly work.

Also, you need to check the site for all utility lines, both overhead and buried. These include electric, telephone, internet, water, sewer, and gas. Make sure you have room to work around and between these utility lines. Not only is it disrupting to break or cut one of those lines, it is also dangerous. Do not dig at a site until you have determined that it is clear of all underground pipe lines. Many old lines are not marked clearly and create a dangerous work situation. In addition, do not work near overhead electric lines because they are easily contacted when working with tower components or cranes. Utility lines can be moved and relocated to allow the turbine to be placed in the best location. This becomes an added erection expense and must be considered when estimating the economics of the wind system.

It is suggested that the soil be evaluated for the design of the foundation. Wind turbine manufacturers will provide a suggested foundation or a series of foundations depending on soil type. Just be aware that a sandy soil will require a larger foundation than a heavy loam or clay-type soil. As part of the planning process you may need to do a soils test and even have a local professional engineer review and approve the manufacturer's recommended foundation. Many local zoning codes require soil tests for foundation designs. Taking a shovel and digging down a half meter will give you an idea about your soil conditions.

Finally, take a final look at your site and imagine what it will look like in 5 or 10 years. Can you still get equipment into the site to do maintenance on the wind system? Will those trees be too large or will someone nearby block access to the site? You are planning for the wind turbine to last many years, but you also need to plan for some unplanned maintenance. Think about the future and what might be needed.

REFERENCES

[1] NRG Systems, Inc. Manufacturers of wind measuring equipment. Web site price list, 2011 (www.nrgsystems.com).
[2] Wind Powering America (www.windpoweringamerica.gov).
[3] British Wind Energy Association, RenewableUK, www.bwea.com.small.
[4] Nelson V. Wind Energy, Renewable Energy and the Environment. New York: CRC Press, 2009, pp. 37–38.

[5] International Electrotechnical Commission, International Standard, IEC 61400–2. Wind Turbines—Part 2: Design Requirements for Small Wind Turbines, 2006.

[6] Wegley, H.L., Ramsdell, J.V., Orgill, M.M., and Drake, R.L., A Siting Handbook for Small Wind Energy Conversion Systems. Department of Energy publication PNL–2521 Rev 1, March 1980.

[7] National Wind Coordinating Colloaborative. Wind turbine interactions with birds, bats, and their habitats. www.nationalwind.org; 2010.

CHAPTER TWO

Needs Evaluation

Determining Power Needed to Meet the Owner's Request

Small wind turbines can and have been used for almost any conceivable application. They have been used for mechanical grain grinding, water pumping, cathodic protection, charging batteries, and electrical power generation. The uses are limited only by our imagination, but the critical issue is selecting a turbine that meets the anticipated need. Because small wind turbines range in size from a few watts to approximately 100 kW, the evaluation of the anticipated load or function is necessary. This chapter presents the characteristics of some of the more popular types of applications and a suggested method of determining the needed power to perform that task or tasks. A common use of small wind systems is to generate utility grade electricity with a provision for the wind system to work interactively with the utility. This is what the industry calls grid-connected wind generation.

GRID-CONNECTED ELECTRIC POWER GENERATION

Small wind turbines of all sizes can be configured to provide grid-connected electricity either by direct connection or through an inverter. Examples of systems that have a direct connection are turbines that use a synchronous or asynchronous generator. The operation is near constant speed (rpm) of the wind turbine rotor, thus reaching its maximum efficiency at a single wind speed. These wind turbines usually have some type of rotor speed control through either blade pitch control or stall–regulated airfoil design. Machines with induction generators (asynchronous) use the excitation (frequency) from the utility to control rotor speed. The output from the machine produces near constant frequency and voltage. A typical machine provides electrical power at 230 to 245 V at a frequency of 59 to 61 Hz to an electrical grid operating at 240 V and 60 Hz.

Small Wind
ISBN 978-0-12-385999-0

Systems that use inverters operate at variable rotor speed, which allows the wind rotors to stay within a zone of maximum efficiency at each wind speed. More and more small wind systems are opting for variable speed operation because more energy can be produced from the range of wind speeds in which it operates. This is partly due to the fact that inverters have greatly improved technically since the early 2000s. For the wind turbine owner, it is a matter of cost of energy per unit of energy produced. Currently, machines less than 25 kW use variable speed generators and larger turbines use constant or near-constant speed generators. More details on the types of generating systems are discussed in the wind turbine components chapter.

The most common use of grid-connected wind machines is for general home or business use where all types of electrical loads are on the systems. In this case the wind turbine is usually connected to the grid between the metering point and the house or business (Figure 2.1). Connected in this manner, all wind power is used by the local loads as long as they exceed the output of the wind turbine and only the excess is fed back into the electric grid [1]. Typically, all capacity is used during the daytime and evening, with the excess fed back during the period from midnight to breakfast time. Exceptions are times when strong storm fronts move through the area, creating periods of excess power generation.

Occasionally a wind machine is added at the location of a single load. The single load is usually some type of motor load or heating load. Again the wind machine is connected to the grid between the metering point

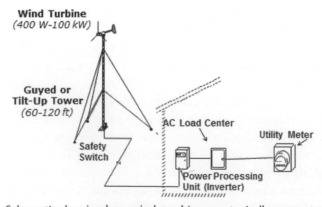

Figure 2.1 Schematic showing how wind machines are typically connected in homes.

Figure 2.2 Example of a wind turbine used for a single load.

and the load. In this case, the wind turbine output goes directly to the load when it is operating or 100% of the wind power goes back to the grid when the load is off. Water pumping and other farm type loads are examples of a single machine on a dedicated load [2]. Figure 2.2 is an example of a wind turbine on a dedicated pumping load. Rarely does one find a wind machine wired to switch between two or more dedicated applications.

STAND-ALONE APPLICATIONS (NO GRID)

Small wind systems are often used to provide power in remote locations where there is no electrical grid. These units are used commonly to charge batteries, which in turn are used to power the loads. These systems were common in the first half of the 20th century before electric utility grids were extended to rural areas. The loads can be powered either by direct current (DC) or through an inverter to power alternating (AC) loads (Figure 2.3). Typical loads for these installations are usually lights and small

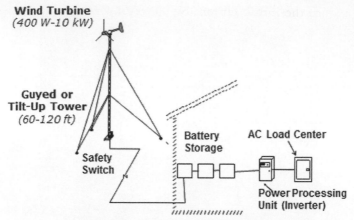

Figure 2.3 Percentages of energy used for various home activities.

appliances. These loads are sometimes hard to estimate because not all of them are used at the same time. One way to estimate the size of wind turbine needed is to consider some small appliances that might be used in the home. If you want to use a hair dryer (1200 watts) and a light (60 watts) together, consider the total wattage (1260 watts) required for operating these items. As you can easily see, the load adds up pretty fast.

Also remember that the load is powered by a combination of the wind turbine and the storage system (batteries). A larger battery storage system allows more items to be used for longer periods of time. However, the storage system needs to be sized to the wind turbine so that it can be fully charged or recharged in a reasonable amount of time; that is, overnight. This is why a wind machine for stand-alone applications will need to be larger than one that is connected to the utility grid.

Other applications for stand-alone applications are for water pumping and other dedicated loads. In these cases, the wind turbine has to be sized large enough to start the application and then keep it running. Applications such as water pumping require about two to three times more power to start the pump than to keep it operating once running [3]. Typically, it is recommended that the wind machine be twice the capacity of a pump motor when it is used in a stand-alone application. Therefore, it is important that you know the power requirements of any individual loads you plan to power when the wind machine is used in a stand-alone application. It is best to consider a low-voltage starting system for motors used in a stand-alone wind system. These starting systems reduce the inrush current by 50 to 70 percent and ensure a safe start each time.

MEASURING CURRENT ENERGY USE

Recently, more small wind machines were sold for grid–connected applications than for stand-alone; therefore, homes and businesses that are considering using small wind machines may already be occupied and are using electrical energy for most applications. If this is the case, it is fairly easy to determine the current energy use because most utilities maintain a 2–year file of past electrical use and can provide that upon request. A typical 2–year energy use would look like the one in Table 2.1. This contains both electrical and gas usage for the past 24 months [4]. This is for a 1,600-ft^2 house in the Midwest where much of the energy load is space and water heating. The yearly electrical energy use is almost 16,000 kWh or approximately 1,325 kWh per month. However, if you want to include space heating and cooling when a wind machine is installed, then the energy use will need to be increased.

Information about home energy use is helpful in separating the various loads into categories that help in estimating what will be powered by the proposed wind system and what loads will be supplied by other energy sources, such as bottled gas or wood for heating. Figure 2.4 shows the percentages of energy used for various home loads. Note that over half of the energy used is for water heating (12%) and space heating/cooling (43%). So, considering the example given earlier, the 16,000 kWh per year will need to be doubled if space heating/cooling is included in the energy load when a wind machine is added. It is important to include all anticipated energy loads when estimating loads for the new wind machine.

Table 2.2 has four years of electrical use for a 2,900-ft^2 southwestern home with a dual–fueled heat pump for space heating/cooling. The dual–fueled heat pump system operates fully on electricity as long as the air temperature is above 35°F, but switches to gas when the temperature is below 35°F. This electrical use includes some heating during the winter months. With a longer record of data, it is easier to see the variations from year to year that occur because of differences in temperature and family activities. In this 4–year period, high and low years varied by approximately 2,000 kWh or 14%. Only one year in four was close to the average [5]. In this case, if all heating/cooling was converted to electric, the 14,600 kWh needed might be increased to 22,000 kWh instead of doubling to 29,200 kWh.

Table 2.1 Billing history from utility for home with gas heating

Bill date	Total electricity used (kWh)	Total electricity charges ($)	Total gas used (therms)	Total gas charges ($)	Total charges ($)
2/2/2012	997	139.74	79	65.32	205.06
1/4/2012	1,176	163.97	76	66.55	230.52
12/2/2011	897	127.27	49	48.01	175.28
11/1/2011	1,038	152.16	16	19.53	171.69
10/3/2011	1,321	191.49	10	16.42	207.91
9/1/2011	1,784	253.77	12	16.14	269.91
8/3/2011	1,833	260.45	12	17.12	277.57
7/5/2011	1,599	229.36	16	21.54	250.90
6/3/2011	1,163	169.71	23	25.31	195.02
5/4/2011	1,411	192.60	52	48.87	241.47
4/4/2011	906	127.69	53	49.79	177.48
3/4/2011	1,300	177.86	77	67.63	245.49
2/3/2011	1,163	160.25	90	83.03	243.28
1/5/2011	1,361	179.07	91	84.53	263.60
12/3/2010	915	123.20	52	53.68	176.88
11/2/2010	959	134.31	18	21.08	155.39
10/4/2010	1,584	215.46	14	18.84	234.30
9/2/2010	1,950	261.71	12	17.20	278.91
8/4/2010	2,028	271.75	14	19.97	291.72
7/6/2010	1,741	235.64	16	21.99	257.63
6/4/2010	1,392	190.46	27	27.99	218.45
5/5/2010	1,121	148.12	34	37.10	185.22
4/6/2010	1,059	139.54	52	51.10	190.64
3/5/2010	1,205	155.38	78	75.25	230.63

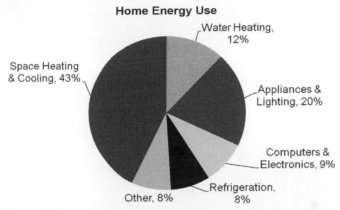

Figure 2.4 Percentages of energy used for various home activities.

Table 2.2 Monthly electrical usage for home with dual-fueled heat pump (kWh)

Month	2008	2009	2010	2011	Average
Jan	1,608	1,578	1,411	1,383	1,495
Feb	1,351	1,351	1,194	1,156	1,263
Mar	1,229	998	1,210	1,077	1,129
Apr	1,300	1,035	988	891	1,054
May	801	749	870	860	820
Jun	1,408	850	974	1,082	1,079
Jul	2,121	1,423	1,429	1,808	1,695
Aug	1,241	1,678	1,563	1,989	1,618
Sep	1,329	1,083	1,345	1,528	1,321
Oct	1,202	798	820	618	860
Nov	1,222	1,019	946	1,057	1,061
Dec	1,009	1,272	1,323	1,256	1,215
Total	15,821	13,834	14,073	14,705	14,608

A third example is a smaller (1,200 ft^2) home that is all electric located in a mountain resort area with seasonal use (Table 2.3). It is used daily in the summer months, but only on weekends in the winter months [6]. These data clearly show the effects of the high heating load in winter, with usage being almost three times as much per month. If a wind machine is selected to meet the largest winter demand, then there could be large amounts of excess energy produced in the summer months. In this case, a careful month–by–month analysis needs to be done to see how well the load demand meets the wind resource. The final selection may reduce the excess in summer and increase the energy purchased in winter as the most optimum

Table 2.3 Monthly electrical usage for all-electric home (kWh)

Month	2008	2009	2010	2011	Average
Jan	2,670	2,630	2,458	1,563	2,330
Feb	1,713	1,555	1,268	1,606	1,536
Mar	1,415	1,590	1,551	1,589	1,536
Apr	706	565	487	304	516
May	424	156	283	286	287
Jun	439	584	434	313	443
Jul	363	584	668	290	476
Aug	564	1,066	880	363	718
Sep	218	623	198	497	384
Oct	573	327	509	540	487
Nov	942	1,137	965	1,293	1,084
Dec	1,103	1,056	865	1,247	1,068
Total	11,130	11,873	10,566	9,891	10,865

operating match. As can be seen in these three examples, each situation is completely different and needs to be evaluated carefully. This can only be done by searching past records of use and looking at both yearly and monthly data.

Businesses and schools would use this same approach to determine the amount of energy needed to satisfy the current energy demands. If businesses operate large amounts of electrical machinery during normal working hours, but not at night, they may want to look at daily energy use patterns. Many locations have higher wind speeds during midday than late evening times, thus making it more attractive to locations with cyclic daily loads. To determine daily cyclic loads may require the present metering system to be updated with one of the "smart" meters that record usages over short time periods such as hourly or every 15 minutes. Peak usages can be measured for each time increment during the day. Matching these short-time use numbers to hourly wind speed data requires more extensive wind data and more data analyses to make a wind turbine selection.

ESTIMATING LOADS FOR NEW CONSTRUCTION

Often it is necessary to determine the energy consumption for a new building or location. A listing of all energy-consuming appliances and loads will be needed and then some estimation of their hours of operation. Many new appliances such as stoves, refrigerators, and washing machines have a label indicating monthly or annual electrical usage. These numbers can be used as a starting point to estimate annual or monthly energy consumption. As shown from the examples of the energy use from existing homes, no two homes are alike, so be careful in using any national average energy use number for your new construction. It needs to be localized for your conditions and living style.

REFERENCES

[1] Clark RN. Wind conversion equipment and utility interties. Lincoln, NE: Proc of Great Plains Agricultural Council Solar and Wind Systems Conference; 1983. p. 83–93.
[2] Clark RN, Vick BD. Determining the proper motor size for two wind turbines used in water pumping. ASME Wind Energy 1995. SED vol. 16, 1995; p. 65–72.
[3] Vosper FC, Clark RN. Autonomous wind-generated electricity for induction motors. Trans. ASME. J Solar Energy Engr 1989; 110: 198–201.
[4] Data from records at Wisconsin Public Service, 2012.
[5] Data from records at Xcel Energy, Amarillo, TX, 2012.
[6] Data from records at Kit Carson Electric Cooperative, Taos, NM, 2012.

CHAPTER THREE

Wind Turbine Components and Descriptions: Essential Components to Have a Productive, Reliable, and Safe Wind Machine

It must be remembered that a wind machine is a machine just like a car, tractor, refrigerator, or copy machine; it has many parts; and there are many different choices where each is designed to meet the various needs and desires of the customers. This chapter first discusses some of the basic parts of a small wind machine and then focuses on a machine used to generate electricity. Machines for other applications are discussed in a later section about applications. Figure 3.1 shows a small wind machine with some major components identified. All machines need a supporting tower to elevate themselves up into the free wind stream. Machines have been installed on all types of towers and towers of all heights. One common type is the pole tower, which consists a single monopole structure that can be either guyed or freestanding. A second common type is the lattice tower, which usually has three or four legs, with braces between the legs. Again these towers can be guyed or freestanding. Manufacturers will suggest a tower for their machine, but will often offer several options for towers from which the owner can choose. On top of the tower will be the nacelle or the enclosure that houses the generator/alternator and any associated equipment needed to control and operate the wind machine. The nacelle offers protection of the electronic components and provides an aerodynamic shape to allow the wind to flow past the main body of the machine.

The tail is used to keep the wind machine facing into the wind and to provide a mechanism for speed control in extremely high winds. Not all machines have a tail because the rotor may rotate around a vertical shaft instead of a horizontal shaft. Other machines may use a drive motor on the tower to keep the machine facing into the wind. A discussion of upwind versus downwind rotors will follow. Finally, the other major component shown is

Small Wind
ISBN 978-0-12-385999-0

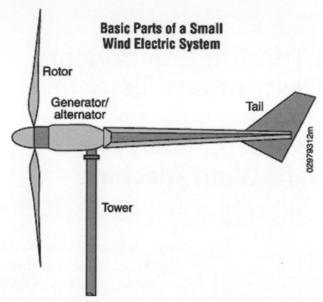

Figure 3.1 Drawing of a small wind machine showing the basic components.

the rotor. The rotor is the heart and soul of the wind machine because it is the component that harvests the wind and converts the kinetic energy in the wind into mechanical power that can be used to generate electricity. Also, the shape and location of the rotor defines the type of machine and how it operates.

The topology of wind machines needs to be discussed in order to understand the various concepts incorporated into the various designs. Two general concepts are discussed that people use to distinguish between basic wind machines. One concept is a division based on the physical position of the main rotating shaft with respect to the earth's surface, and the other has to do with the type of rotor design and the way the rotor harvests the wind.

HORIZONTAL AXIS AND VERTICAL AXIS MACHINES

A wind machine is often described by the position of the main rotating shaft with respect to the earth's surface. Therefore, a horizontal axis wind machine has a shaft that is parallel to the earth's surface. The rotation of the rotor is perpendicular to the air flow across the earth's surface as shown by the machines in Figure 3.2. A horizontal axis wind machine requires some type of device to keep the rotor into the wind. Many small wind machines use a tail to keep the rotor into the wind, and larger machines use a drive motor and wind

Figure 3.2 Horizontal axis wind turbines.

direction sensor to move the rotor so that it stays across the wind. Because horizontal axis machines are placed easily on towers of varying heights with little design modification, they are widely used for many applications. The acronym HAWT is used for horizontal axis wind machines or turbines.

A vertical axis wind machine has its main rotating shaft perpendicular to the earth's surface or vertical with respect to the earth surface as shown by the machines in Figure 3.3. A main advantage of a vertical axis wind machine is that it accepts wind from any direction—the turbine does not have to rotate to face the wind. However, because of their shape, many vertical designs require a higher starting wind speed than horizontal types.

Figure 3.3 Vertical axis wind turbines.

Another advantage of the vertical axis design is that all the gearboxes, generators, brakes, and so on can be located near or on the ground, which makes maintenance much easier. A significant disadvantage is that vertical axis wind machines cannot be placed easily on taller towers without major design modifications or expensive towers. The acronym VAWT is used for vertical axis wind machine or turbines.

DRAG DEVICES

Wind machines are also classified according to the way that they capture the energy in the wind. Some machines act as wind catchers in that the wind is caught in a sail or a cup or it pushes against a plate. These machines are classified as drag devices because the wind pushes or drags the blades around an axis (see Figure 3.4). Many drag devices are also vertical axis machines because it is easier to design a machine to rotate around a vertical shaft. A good example of a drag-type wind machine is a cup anemometer. The wind pushes the cups around, and the rotating shaft is connected to a small generator creating a small electrical signal. Sometimes shields are built on the side of the machine where the blades come against the wind to reduce the drag on the rotor as it moves against the incoming wind. In all cases, the rotor does not turn faster than the wind speed, creating a machine with relatively low rotational speeds. Many old historical wind machines were some form of a drag-type wind machine. All across the Middle East and around the Mediterranean Sea, old wind machines

Figure 3.4 Drag-type wind machines.

Figure 3.5 European wind machines using sails on wooden frames. These were used for water pumping and grinding grain.

with sails were a drag-type machine used for water pumping and grinding grain. The Dutch wind machines also used sails to capture the wind and had a slow rotation used for water pumping (see Figure 3.5).

Typically, drag-type wind machines require large rotors to obtain the desired power. Their efficiencies are low because of the slow operating speeds and heavy weight. Typically, efficiencies are less than 15%. These machines are popular with inventors and home builders because almost any device that will rotate and catch the wind will produce some power. However, they have not proven cost-effective for producing electricity.

In the 1930s, S. J. Savonius developed an "S rotor" or, more appropriately, the "Savonius" wind machine, which showed improved efficiencies over most drag-type wind machines. However, most people do not follow the original design and end up with simply another drag device. Figure 3.6 shows two S rotors, one with both blades connected to form a solid blade and the other with an opening in the middle to allow some air to flow through the center. Savonius's original design is the one with the

Closed

Open

Figure 3.6 Schematic of a closed and open Savonius rotor.

Figure 3.7 Two different types of Savonius rotors. It is difficult to scale up to larger than 5 kW.

opening [1]. Tests conducted show that the split Savonius has efficiencies approaching 30% as compared to 15 to 18% for the closed Savonius. There has been great difficulty in scaling up the Savonius like the one shown on the left in Figure 3.7, which lasted only a few days under prototype testing. The greatest use of Savonius rotors is for powering navigational buoys as shown in Figure 3.8.

LIFT DEVICES

Wind machines classified as lift devices use blades or wings that have an aerodynamic shape similar to airplane wings. The aerodynamic principles by which they are driven can best be explained by studying the diagram in Figure 3.9. As the airfoil moves through the wind, it is subject to lift-and-drag forces. The lift-and-drag forces are usually determined experimentally for airfoils as a function of the angle of attack. The angle of attack is the angle between the relative wind and the chord line of the airfoil. The relative wind, sometimes called the apparent wind, is composed of the vector sum of the wind across the ground and the motion of the rotor blade. The horizontal component of the lift on the blades makes the rotor turn about the axis. The angle of attack controls the amount of lift that contributes to rotation. Power output increases as the speed of the rotor increases as the

Figure 3.8 Open top Savonius-type rotor used to power navigational buoys. Small battery-charging applications.

wind speed increases and the angle of attack remains constant. Fixed pitch machines are where that angle of attack is fixed and does not change during operation at all wind speeds. Turbines equipped with variable pitch controllers have the capability of varying the angle of attack to control rotor speed and power output. More will be said about pitch-controlled machines in the discussion about controlling rotor speed and over-speed control.

Lift-type wind machines typically have rotor blades that are thin and operate at speeds much faster than the wind. One measure of wind turbine rotor performance is the tip-speed ratio, which is the speed of the tip of the rotor blade divided by the wind speed. Figure 3.10 shows typical performances of five different general types of wind machines. This chart was first published in 1975 in a report for the National Science Foundation [2]. Additional research has shown that by using specially designed airfoils, the performance of high-speed rotors can be improved slightly to approach 50% and the performance of Darrieus rotors to 45%. The tip-speed ratio point of maximum efficiency for a high-speed rotor blade is about six to seven for a lift device and about 0.5 for a drag device.

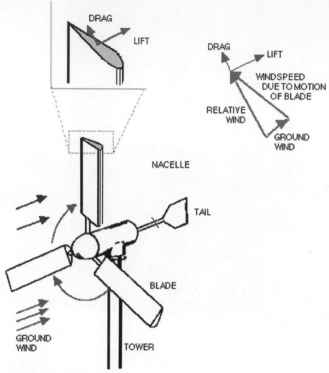

Figure 3.9 Schematic of a blade showing the lift and drag forces imposed on it as the blade rotates through the wind.

In addition to operating at a higher speed, lift devices are always lighter and cost less to construct. The rotor on a lift device can produce as much as 50 times more power per unit area of blade than a drag device. The most common type of lift device is a three-bladed horizontal axis wind machine. The number of blades is not as important for aerodynamic performance as it is for stability and maintaining the rotor across the wind. Wind machines with one blade and a counter weight have produced almost as much power per unit area as machines with two or three blades. Once blade numbers get above five, the blades tend to interfere with each other and the individual blades no longer see new clear wind flows. Additional blades do not improve performance or stability, only the cost. Most arguments about blade numbers revolve over whether it should be two or three. Three blades give the rotor stability at all points during the rotation whereas there are four points during a single revolution where a two-bladed rotor goes through a zone of unstableness. When a rotor blade is exactly horizontal, there can be an upward or downward thrust

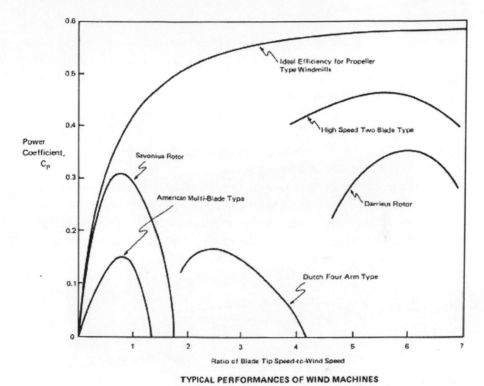

TYPICAL PERFORMANCES OF WIND MACHINES

Figure 3.10 Efficiencies of different wind turbine rotors as compared to the ideal efficiency for propellers. Developed by Frank R. Eldridge in 1975.

on the rotor hub, which creates a twisting motion on the hub. Likewise, when the rotor is exactly vertical, there is a twisting motion right or left that causes loading on the yawing motion of the entire nacelle. Strengthening the hub and yaw system to overcome these rotor loads and additional fatigue requires additional costs. Often, it is simply less expensive and easier to add a third blade. A comparison of the number of wind machines with three blades versus two blades indicates that the majority of manufacturers agree that three blades are better and less costly to maintain.

ROTOR DESIGNS

Horizontal Axis

The first rotor design focused on here has three blades mounted on a horizontal axis, which is the most common rotor design found on small machines. Many of these small wind machines use a single airfoil design for the entire length of the blade, but it is common for the blade to be tapered so that it is larger at the root and smaller at the outer end. This is done to support the maximum bending moment near the hub and to minimize the weight of the rotor. These blades are sometimes also twisted so that the blade will have the same angle of attack throughout the entire length of the blade. On machines less than 3 kW, the blades are made flexible enough to twist with increased speed and loading. This provides for the blade to operate at maximum performance at all wind speeds.

For blades five meters or longer, you often see a family of airfoils used rather than a single airfoil. With these longer blades, the air flow experienced by the rotor is much different near the root and out near the tip. Figure 3.11 shows an airfoil series and their placement along a 5-meter blade. The

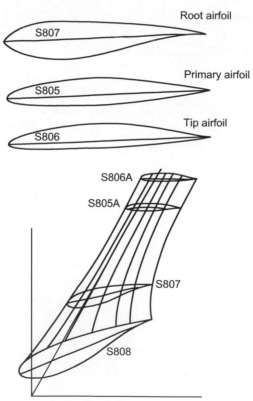

Figure 3.11 Layout drawing of a 5-m blade showing how different airfoil shapes are used to construct a complete, efficient blade [3].

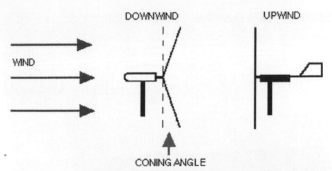

Figure 3.12 Schematic of a downwind and upwind rotor configuration for small wind turbines.

primary airfoil is used for the outer 30 to 40% of the blade [3]. The root airfoil is used for about 25 to 30% of the blade near the root. The middle section is a transition airfoil that makes a smooth surface from root to primary airfoil. It can easily be seen why larger rotor blades become sophisticated and require careful design and construction—even blades for machines in the small classification.

Blades can be located upwind of the tower or downwind of the tower (Figure 3.12). Upwind rotors less than 50 square meters often use a tail vane to keep the rotor turned into the wind. Machines larger than 50 square meters use a small servo motor and gearing system to keep the rotor facing the wind. Machines with motor drives are usually referred to as machines with positive yaw or active yaw. Yaw is the motion of the nacelle (generator and rotor hub) about the tower top. The design of the tail has to be long and large enough to overcome the gyroscopic loads that want to make the rotor turn out of the winds.

Blade Materials

Wind machine rotor blades have been constructed of wood for centuries. The first successful blades constructed of metal were for the farm windmill. These blades were curved metal and were mounted on rings that made up the rotor. Again the first blades using airfoils were constructed of wood. Jacobs wind machines used spruce wood for their blades beginning in the 1930s. A photo of an all–wood rotor airfoil is shown in Figure 3.13. Several early wind machines built in the 1970s and 1980s attempted to use metal blades constructed like airplane wings. All of these failed because of metal fatigue at the root. Other early blade materials were foam filled with a fiberglass skin like the one shown in Figure 3.14. These blades were either

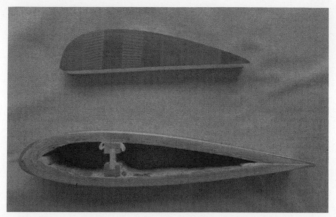

Figure 3.13 Airfoils constructed from wood. The top blade section is all wood with no spaces, and the lower blade section is a wood laminate with air spaces and a rib brace.

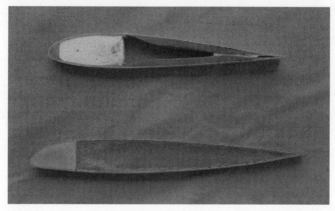

Figure 3.14 Blade cross sections of fiberglass blades with foam filling. Both of these blades have metal placed in the leading edge to add weight to help the blade rotate.

too flexible and had poor performance or were too stiff and cracked due to fatigue similar to the metal blades.

Now many blades less than 3 meters in length use an injection mold with carbon fiber filaments for strength as shown in Figure 3.15. The amount of carbon fiber is adjusted to make the blades rigid enough to perform well at normal wind speeds, but still flexible enough to bend in high winds without breaking. They work well on small machines with fixed pitch and rigid hubs. Medium-sized blades (2 to 4 meters) are manufactured using a pultruded fiber–reinforced polymer process. These blades are more rigid

Figure 3.15 Injection molding is the newest method of constructing high-strength, flexible rotor blades.

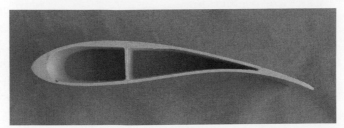

Figure 3.16 Cross section of a pultruded blade when the material is pulled through the mold.

than injection molded blades and are usually not tapered. However, it is possible to have a more detailed airfoil as shown in Figure 3.16.

Larger blades are normally made of fiberglass layup construction. These blades usually have a more complex airfoil shape and require a split mold for construction. Also, because of the length, an internal brace or spar is included in the blade as shown in Figure 3.17. This construction is almost identical to the construction process of the larger blades for megawatt machines.

Note that the airfoil shape of all these blades is curved on the top and almost flat on the bottom. This is the shape of the airfoils used on horizontal

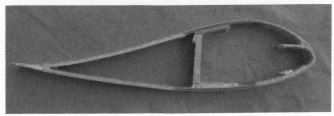

Figure 3.17 Fiberglass blade with carbon fiber fill at the spar brace for carrying stresses and loads. This is used to construct larger blades that are much lighter than other materials.

axis wind machines. Airfoils have gone through several iterations since the mid–1980s, and almost all manufacturers use airfoils that have been designed especially for wind machines. Many older machines are still operating with older type airfoils, which have proven to be less efficient; however, they have already paid for themselves, and will still produce significant power if maintained. Some of these machines are being refurbished and resold into the small machine market. These old blades need to be inspected closely for cracks and separation along the leading and trailing edges where the two sections were seamed together.

Vertical Axis

Airfoils for vertical axis wind machines are normally symmetrical in shape, meaning that both upper and lower surfaces have the same shape. A significant difference between a vertical axis rotor and a horizontal axis rotor is the time that the airfoil is exposed to direct wind. A vertical axis rotor sees actual wind for about one-fourth of the rotation, and the horizontal axis rotor is always in direct wind. Using an asymmetrical rotor, the airfoil produces some lift on the backward part of the rotation [4]. This improves the performance of vertical axis machines greatly. Figure 3.18 contains some symmetrical airfoils for vertical axis wind turbines. These airfoils are used mostly on either Darrieus or gyromill vertical axis machines. Some gyromill designs are now using slanted blades instead of exactly vertical ones. This gives a more aesthetically pleasing look, but does not improve performance greatly. Many vertical axis wind machines do not use a true airfoil but rather use a design that is mostly a drag device that catches the wind.

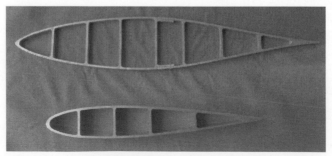

Figure 3.18 Symmetrical airfoils used for vertical axis wind turbines. These are extruded aluminum, but new ones are often fiberglass or pultruded.

ROTOR HUBS

The rotor hub is the component that usually holds the blades and connects them to the main shaft of the wind machine. It is a key component not only because it holds the blades in their proper position for maximum aerodynamic efficiency, it also rotates to drive the generator. Hubs come in many different shapes and configurations, mostly dependent on the type of generator used and the design of the rotor blades. This discussion covers the five most common general types of hubs found on small wind turbines. The first type of hub incorporates a hub with the housing containing the magnets for the permanent magnet alternator. For smaller machines, the rotor blades are bolted directly to a plate welded to the housing or can that holds the magnets (Figure 3.19). This hub also includes a bearing that slips on a fixed

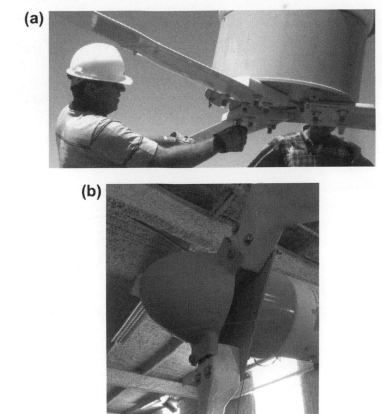

Figure 3.19 (a) Rotor blades are bolted directly to the magnet can of the permanent magnet alternator. (b) Rotor blades are bolted to a plate welded to the magnet can of the permanent magnet alternator.

Figure 3.20 Cast hub for turbines with fixed pitch blades and a main shaft. These designs usually have a gearbox.

shaft. The only rotating part of this wind machine is the hub with blades and magnets all attached. The bearing needs to be large enough in diameter to add stiffness to the system so that wind loading on the rotor will not affect the gap spacing between the magnets and the stator. Blades may be bolted flat to the hub plate or there could be brackets for the blades to slip into and securely fastened. Machines up to about seven meters rotor diameter have been operated successfully with this hub configuration. The second type of hub is attached rigidly to a rotating main shaft that connects to a gearbox to increase the rotational speed up to the operating speed of the generator. This hub is often constructed of cast material to get the right alignment of the rotor blades (Figure 3.20). This hub construction accommodates blades up to several meters in length, well beyond the largest machines classified as small wind systems. Once the blades are attached to this type of hub, they remain at the same pitch and twist throughout all operating conditions. These two types of hubs are found on the majority of small wind machines because of cost and reliable operation.

Some small wind machines include a full blade pitch assembly. This requires a special hub with the capability of changing the pitch of each blade. The more common type includes a motor/gear drive on each blade that moves a cantilever arm that makes the rotor blade twist. It is extremely important that the entire system of blades move at the same time and the same amount. If one blade has a different pitch, it may either be carrying most of the load or not carrying any of the load. Uneven pitching of rotor blades is a major cause of rotor failure on machines, small and large, causing a rotor imbalance resulting in a tower strike with a blade. Typically, only

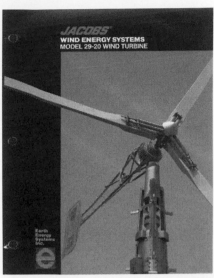

Figure 3.21 Springs attached to a hub allow blades to pitch if rotor speeds become excessive. This is an over speed device. *(Photo of Jacobs literature).*

larger machines have rotors with a fully controlled pitch. However, several small machines use heavy springs to hold the blades at a fixed pitch until an over speed occurs (Figure 3.21). In an over speed condition, the springs will stretch and allow the rotor blades to turn to a lower pitch setting that causes the rotor to lose its lift capacity, thus causing the rotor to slow down. This concept is discussed more in the section on over speed control issues.

The fourth type of hub on horizontal axis wind machines is a teetering hub, which is used when the machine has two blades. The hub has a flex in it that allows the blade passing by the tower to flex because it has less load on it for that instant it is by the tower. Teetering hubs are especially useful on downwind machines because of the changes in load when the blades pass behind the tower. The larger the rotor, the more important it is to have a teetering hub when two blades are used. Keeping the teetering point lubricated has been a problem with many small machines using teetering hubs. Maintenance issues with teetering hubs have caused many designers to change to the three-bladed fixed hub designs.

Hubs for vertical axis machines look very different and are usually not called hubs, but a similar device is used to support the rotating structure. This may look like a bearing housing with mounting brackets attached to hold the support arms for vertical airfoils. In almost all cases these brackets are supporting structures that are in tension because of the centrifugal forces pulling the blades from the center of rotation. On the Darrieus vertical axis design, the lower mounting bracket carries the most centrifugal forces; many blade failures occur at this mounting fixture. The same problem occurs with gyromill–type wind machines.

ENERGY CONVERSION SYSTEMS

The configuration of small wind turbines varies widely because they do not all have the same components in the section commonly called the "drive train" for large wind machines. Typically, a drive train is composed of a main shaft with gearbox, brakes, and generator all attached in some sequential order mounted on a platform (Figure 3.22). For small machines, the gearbox may be cast to include the tower mount with the generator attached or the unit may not have a gearbox and the rotor shaft is connected directly into the generator. There are almost as many drive train configurations as there are small machines manufactured. The drive train components are usually housed in a covering, and the complete unit on top of the tower is called the nacelle. Many small machines can be recognized easily by the shape of their nacelle.

Vertical axis machine manufacturers always indicate that a major advantage of the vertical axis machine is that all the energy conversion components are at or near ground level for easy access when service or repairs are required. Again, the arrangement of gearboxes, generators, and brakes can vary greatly among the different manufacturers of vertical axis machines. A main difference between vertical and horizontal axis machines

Figure 3.22 The drive train of a small wind turbine consists of a rotor hub, disk brake, bearings, gearbox, and generator.

can be the length of the main shaft and the types of bearings used to support the loading from the wind rotor.

Generators

Small wind systems use a variety of generator types, ranging from small direct current (DC) generators/alternators to synchronous and asynchronous generators. Small machines used for battery charging use a generator producing DC power that can be fed directly into batteries using a charge current controller. The Wincharger Company and Jacobs Wind Plants began making a wind machine like this during the 1920s. Thousands of these units were sold through dealers and catalog order stores. The popularity and use of these small battery charging wind systems stopped when the Rural Electric Administration began installing transmission lines and providing utility electricity to rural America in the late 1940s and early 1950s. Small wind systems disappeared almost completely during these years of inexpensive electricity and easy access to centrally generated electricity.

Permanent Magnet Generators (PMA)

Today, many small wind systems are made to either charge batteries or be connected to a utility grid. These units use a permanent magnet alternator type of generator. The PMA uses the same principles of a magnetic field rotating around a wire-wound stator. Rare-earth magnets or magnets made of neodymium–iron–boron are placed in a ring that rotates around the stator. The assembly of magnets is called a magnet can and is often an integral part of the hub in smaller machines (Figure 3.23). The wire-wound stator

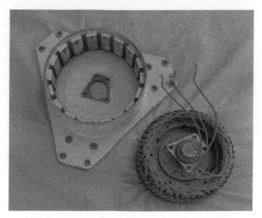

Figure 3.23 Stator and magnet can of a permanent magnet alternator. Note flanges on the magnet can for mounting the rotor blades.

remains stationary, and power is taken from the stator and conducted down the tower either through slip rings at the tower top or through a twist cable. The PMA offers several advantages over the older type of DC generator system. One advantage is that only one-half of the generator has to be wire wound, which makes the system more reliable at a lower cost. But probably the best advantage is that power can be transferred easily to either DC or alternating current (AC) output [5]. Because output voltage and current from a PMA varies with rotational speed, a rectifier is used to make the output suitable for battery charging as DC power, and an inverter is used to make the power suitable for AC systems at a fixed voltage.

Probably the most important component that has made the PMA the most common type of generator used in small wind machines is the low-cost, reliable inverter that has been developed since the 1990s. The use of new improved digital electronics has allowed for inverter technology to be vastly different from than just a few years ago. Both wind and solar electric systems have benefited from improvements in inverters, especially in better reliability and reduced costs. The availability of inverters from third-party vendors has compelled some wind turbine manufacturers to adjust the output of their machines to match a particular size of inverter. An example is changing a 1.5 kW wind machine to a 1 kW wind machine to match a 1 kW inverter or changing a 2 kW wind machine to a 3 kW machine to match a 3 kW inverter. This optimizes the system for maximum efficiency.

Induction Generators

Induction generators are used mostly on medium-sized small wind machines; that is, those that are from 5 to 65 kW in size. Induction generators are used because they are made in large numbers, are relatively inexpensive, have low operation and maintenance costs, and require simple controls to manage the voltage and frequency match to the utility grid. An induction generator is simply an induction motor driven above its synchronous speed. The synchronous speed of induction motors is typically 1,200, 1,800, and 3,600 rpm, depending on the number of poles in the generator windings. An induction motor/generator does have some internal slip (2 to 5%) due to its construction; therefore, it is classified as an asynchronous motor/generator. The amount of slip is actually dependent on the torque applied to the generator. The 1,800 rpm generator is the most commonly used speed, thus requiring a gearbox to increase the wind rotor speed up to the operating speed of the generator (Figure 3.24). A major disadvantage of the induction generator is that it controls the speed of the

Figure 3.24 A small wind turbine with an induction generator. Almost all machines with induction generators have a gearbox.

rotor to a near constant rotational speed and the wind rotor does not operate at its peak efficiency but over a very narrow range of wind speeds, normally about a 1 to 2 m/s range.

An induction generator requires that the unit be connected to the electrical grid at all times. The voltage, frequency, and reactive power for excitation are drawn from the utility. The amount of reactive power drawn by induction generators of the size used in small machines is almost immeasurable and can be offset easily by capacitors [6]. At some installations above 50 kW, the utility will require some corrective measures for the consumed reactive power, but this is not very common. An induction generator is almost like a plug-and-play type of interface with the utility. It certainly eliminates the need for inverters or other power conditioning equipment. It also has a built-in safety function because if the utility line fails, then the turbine stops generating power and the line repairmen are protected from any feedback from the wind turbine. The wind system control system must provide for a suitable shutdown procedure in case of loss of electric utility power. This procedure is usually covered under the emergency shutdown plan.

Wind turbine startup with an induction generator can be accomplished in several different ways. One common way is to measure the wind speed; once the wind speed reaches the desired level for the wind machine to be producing power, the controller releases the brakes and engages the induction generator with the utility. The generator operates as a motor and spins the wind turbine up to operating speed. Once it reaches operating speed, the wind starts trying to spin the rotor faster and the motor/generator changes from motor to generator [7]. Because this operation usually occurs in less than a minute, the unit does not consume much power during start up. The wind machine typically operates more time as a motor at shutdown in low winds than it does at startup. The startup wind speed may be set at 5 m/s and the shutdown wind speed at 4 m/s to avoid the machine starting and stopping too rapidly. Another common way that units are started is that they are allowed to free wheel (turn slowly) in light winds—when the rotational speed reaches the synchronous speed, the generator is engaged to the utility. This plan eliminates the consumption of power on start up, but can be a problem when winds increase rapidly due to weather changes. Once the turbine begins turning and approaches the operating speed of the induction generator, it is accelerating rapidly and can pass through the operating range easily in a matter of seconds. The generator may not speed up fast enough to catch up with the wind rotor. This is a rare occurrence, but it does happen with some controller applications. These control strategies are used on either horizontal or vertical axis machines with induction generators.

Synchronous Generators

Synchronous generators are not normally used on small wind machines because of the additional equipment needed to make them operate efficiently. One advantage of a synchronous generator is that it can produce utility-type power without an external power grid. The unit must provide its own voltage and frequency control and not depend on the utility. Additionally, most synchronous generators are used with machines that have rotor blades with variable pitch control. The speed of the rotor is controlled by changing the pitch of the blades, which helps provide a constant input speed. Several megawatt wind machines use this system, but almost no small machines.

Gearboxes

Because the rotational speed of the wind rotor is much slower than the needed rotational speed of induction generators and most synchronous generators, some type of speed increaser is needed. Therefore, gearboxes or speed increasers are placed in the drive train assembly. Initially, many designers used off-the-shelf gearboxes designed as speed reducers and ran them backward. This accomplished the needed effect and demonstrated the application of the gearbox, but reliability was poor and the gear ratio was not always optimum for the rotor diameter. Gear manufacturers soon began making gearboxes designed as speed increasers and would design units at

POWER CURVES
Enertech 44

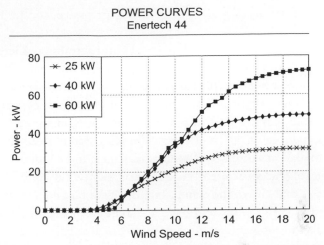

Figure 3.25 Measured power curves for three rotor speeds using different gear ratios in the gearbox. Power was corrected to standard density, and the wind speed was measured at a 25 m hub height [7].

needed ratios. Gear ratios for machines with induction generators range from 5:1 to 10:1 for small machines less than 1 kW and are approximately 35:1 for 50 kW size machines. Figure 3.25 shows the effect of gearbox ratio on the performance of a 50 kW wind turbine with an induction generator. The shape of each power curve is changed, and the maximum power obtained is also different. Because the gearbox ratio controls the speed of the rotor when an induction generator is used, selection of an optimum ratio is extremely important [7].

Gearboxes used in wind turbines are generally of either parallel shaft construction or planetary gear construction. For small machines, parallel shaft-type gearboxes are more popular because of lower cost and quieter operation. Planetary gearboxes are usually smaller in size, but make more noise than the parallel shaft-type construction. Sometimes manufacturers will use a combination of the two types if the gear ratio is large and space is an issue. Also, some manufacturers have opted to custom design the gearbox case to be an integrated design to include yaw bearings and a generator mount. A problem with the integrated gearbox design is that if the gearbox needs repair, then the whole unit has to be removed from the tower.

Gearboxes are not commonly used with PMA or DC generators. This is an advantage of using these types of generators over an induction generator. It also reduces the weight and size of the drive train assembly.

BRAKES

A braking system is required on small wind systems in order to control the rotor in abnormal conditions. These conditions may range from loss of load (electrical line outage), failure of a control system component, or need for scheduled maintenance. The operation of brakes can be grouped into three operating conditions: normal, normal stopping, and emergency stopping. Sometimes a different type of braking system is needed for each condition. Brakes are often used in normal generating conditions. They are used to control the rotor speed in excessive high wind conditions, which may occur less than 5% of the time. Machines with tails used for keeping the rotor turned into the wind will have a system that forces the rotor out of direct wind and causes the rotor to slow down. The practice of turning the rotor partially out of the wind is called furling. The rotor can be turned to the side or lifted vertically when the rotor reaches a maximum rpm. Pitching the blades so that they do not produce full power is also a form of braking to control rotor speed. Lightly applying a mechanical brake is not an acceptable method of controlling the rotor because the amount of pressure applied cannot be controlled remotely.

In addition to speed control, another condition that requires braking is normal stopping and starting. Some wind systems with larger rotors use a control system that applies a small brake to hold the rotor when no power is being produced. A parking brake is often placed on the high-speed shaft of systems with gearboxes because the brake can be smaller and not require as much pressure to hold the rotor in place. Parking brakes are also used to hold the rotor in place during maintenance and repair activities. The parking brake is not applied until the rotor is below 5 rpm.

The third condition where a brake is needed is during emergency situations and there is an immediate need to stop the rotor. Emergency brakes must be able to stop the rotor is less than 10 seconds. They must be large and mounted on the main shaft because of the large amounts of torque that must be applied to stop the rotor. Many wind turbines in the 40 to 80 kW size experienced failure because the gearbox failed when the brake, which was mounted on the high-speed shaft, was engaged.

Mechanical Brakes

Today, most mechanical brakes are disk-type brakes that are similar to brakes on a car. A large disk is mounted in the shaft, and a caliper with pads is

Figure 3.26 Disk brake and caliper for an 80 kW wind turbine with an induction generator. The brake is on the low-speed main shaft.

mounted over the disk (Figure 3.26). It is recommended that brakes be operated in a "fail-safe" mode, which means that the brake pads are spring loaded and are always applied when no power is supplied to the brake. If any type of failure occurs during operation, the brake will be applied and the turbine will be stopped to prevent a "runaway." Mechanical brakes of this type can be powered or activated by electrical switches, hydraulic pressure, or pneumatic pressure. Adding a hydraulic system with a pump and storage tank adds considerable weight and bulk to the wind turbine nacelle. It is the same when an air compressor is added for pneumatic brakes. Also, these pressure systems provide for many potential maintenance problems because it adds another motor and compressor/pump to maintain and all the piping creates opportunities for leaks. An electrical actuated disk brake system works best for small wind systems.

Another type of mechanical brake that has been used is a rotating disk brake or "Stearns" brake (Figure 3.27). This type of brake is almost always used on the high–speed shaft and is used as a parking brake. It consists of usually three disks that rotate with the shaft and has spacers or friction plates between the disks. When the spring is applied, it compresses the disks and plates together to hold the shaft in place, preventing rotation [8]. When the spring is released, the disks are free to turn. Again, the preferred method of configuring this brake is with a spring released when there is no power to the brake so that the shaft will not turn unless the brake is powered up. These brakes have been also used as primary brakes, but many gearbox failures have left these brakes useless.

Figure 3.27 Components of a "Stearns" brake, which is normally mounted on a high-speed shaft when an induction generator is used.

Dynamic Brakes

A dynamic brake consists of a series of resistors and capacitors that are switched in and out of the electric generation circuit. By adding resistance to the electric generator circuit, a greater load is placed on the generator, thus requiring more power than the generator can produce and causing the generator to slow down. Although the rotor will not always stop completely, it will be slowed to a low rpm whereby a parking brake can stop it. Unlike mechanical brakes, this braking system is usually located near the control panel on the ground where it can be checked easily for proper operation and is serviced easily. With proper instrumentation, this brake can be used to regulate the speed of a rotor during normal operation because the resistors can be energized as single resistors or as the whole group. Some small turbines use this type of braking system as a maintenance brake only and offer no other braking system.

Aerodynamic Brakes

Aerodynamic brakes consist of devices that are either attached or embedded into the rotor blades that will deploy to provide extra drag and slow the rotor. The most common type of aerodynamic brakes is tip brakes. Tip brakes are often plates on the end of blades that are mounted perpendicular to the blade on a hinge (Figure 3.28). When the tip brake is deployed, it flops open and the plate drags in the wind flow (Figure 3.29). Another design of a tip brake is, when the end section of the blade is made to swivel and turn 90° so that the

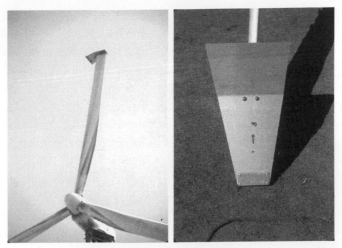

Figure 3.28 Tip brake used in a 50 kW wind turbine for over speed control.

Figure 3.29 Tip brake showing the spring that holds the tip in place until released due to over speed. The brake opens and slows the rotor.

end of the blade is across the plane of rotation. The concept is the same for both types of tip brakes in that a device at the end of the blade creates a large drag which slows the rotor. Tip brakes do not stop the turbine, but slow the rotor and keep it from over speeding in extremely high winds or in the case of a loss of load condition. Normally, tip brakes have to be reset manually by climbing the tower and locking the brakes back into the normal operating position. Some of the newer designs use an electric solenoid to hold the tip brake in place and can be reset by stopping the rotor completely.

Another type of aerodynamic brake is the use of flaps or ailerons. This is where a portion of the blade bends and creates extra drag to slow down the rotating speed. These are similar to flaps seen on airplane winds used during the landing of aircraft. Because of the complexity of the control system and the extra weight associated with ailerons, they are not found commonly on small wind systems. Most designers have found it much simpler and less costly to pitch the entire blade than to add a system to deploy a flap on the outer 20 to 30% of the rotor blade. This is why machines that approach the 100 kW size and larger utilize pitching blades for rotor speed control.

CONTROLLERS

Rapid developments in computer and sensor technology have been incorporated into wind system controllers. There is little resemblance between the controllers used today and those used just 20 years ago. Single-chip computers that could be programmed externally with the program downloaded to the chip were an early advancement. These were some of the first controllers to use real-time data to make operational decisions [9]. Continued development produced a programmable logic controller (PLC) that was rugged enough for the wind turbine environment and fast enough to make appropriate decisions for wind turbine controls. Lastly, systems became small enough to be placed in a package that was easily mountable in a wind turbine.

The wind turbine controller serves several functions. It controls the start-up and shutdown of the wind turbine rotor. It releases the parking brake at the correct time and monitors the rotor speed. At the preselected rotor rpm, the controller engages the electrical generating system or load if some function other than electrical generation is being performed. The controller monitors output to determine if current, voltage, and frequency are within typical operating conditions. It creates a fault condition that causes the turbine to stop if all operating parameters are not within preselected values. The controller may monitor several operating conditions and report the status to a supervisory control and data acquisition (SCADA) system. SCADA system output may be displayed near the wind turbine, such as in the office or home of the owner, or it may be located at the manufacturer's plant. Examples of SCADA systems and their use are discussed in Chapter 11, Operation and Maintenance.

One main function of the controller is to monitor and provide over speed protection for the wind turbine. Many small wind turbines have a passive over speed control that furls the rotor out of the wind when the rotor speed exceeds a selected point. However, a controller may also begin to add additional load to slow the rotor or apply braking to the system. Over speed protection is the most important event that must be controlled because it almost always ends with some type of failure.

Controllers are probably one component that has changed and improved more than any other in the wind turbine system. Controllers now provide real-time monitoring and rapid decision-making changes that ensure a safe operating system that improves reliability and overall performance. The integrated wind machine operates and performs like any completely automated machine such as a refrigerator, central heating and cooling system, or an airplane on autopilot. Because of the ever changing wind conditions, the wind system controller must react instantly to provide a generating system that operates at maximum efficiency.

REFERENCES

[1] Gipe P. Wind Power, Renewable Energy for Home, Farm, and Business. White River Junction, VT: Chelsea Green Publishing; 2004.
[2] Eldridge FR. Wind Machines. National Science Foundation, Division of Advanced Energy and Resources, AER-75-12937 1975.
[3] Nelson V. Wind Energy, Renewable Energy and the Environment. New York: CRC Press; 2009.
[4] Clark RN. Design and initial performance of a 500-kW vertical-axis wind turbine. Trans ASAE 1991;34(3):985-91.
[5] Clark RN. Performance of small wind-electric systems for water pumping: Windpower 94. Am Wind Energy Assoc; 1994. p. 627-634.
[6] Johnson GL. Wind Energy Systems. Englewood Cliffs, NJ: Prentice-Hall; 1985.
[7] Clark RN. Wind electric generator performance during 12 years of operation. Appl Engineer Agric 1995;11(5):745-9.
[8] Clark RN, Ling S. Performance and maintenance experiences with a wind turbine during 20 years of operation. Global Windpower. Am Wind Energy Assoc., CDROM; 2004.
[9] Clark RN, Ling S. A smart controller for wind electric water pumping systems. Houston, TX: 15th ASME Wind Energy Symp; 1996. p. 204-208.

CHAPTER FOUR

Towers and Foundations: Support Structures That Make the Project Successful

Towers are a critical component in a small wind system because the wind turbine has to be placed where it can harvest the wind resource that is available. Towers must be strong enough to support the wind turbine and transfer the loads caused by the wind to the ground. Towers have been built for centuries, and modern towers are subject to the International Building Code, Uniform Building Code, or their local equivalent [1]. The availability of various types and designs of towers for small wind systems has been expanded greatly due to the rapid expansion of cellular phone technology and the demand for many cell phone towers. Therefore, there are several tower options for use with small turbines.

There are two main considerations in selecting a tower that is structurally sound to meet the requirements of a satisfactory tower for any wind turbine. First it has to be able to withstand the forced caused by the wind load. Loading is a function of the wind force on the cross–sectional area of the rotor and any protruding structure outside of the rotor area. For almost all small wind machines, all of the drive train equipment is within the projected cross–sectional area of the rotor. The wind force is a function of air density, wind speed, intercepted area of the wind turbine rotor, and a dimensionless coefficient that represents the shape of the structure.

$$F_w = \tfrac{1}{2}\,\rho A V^2 C_D, \qquad\qquad [4.1]$$

where F_w is the wind force on the tower or drag caused by the tower, ρ is the density of the air in kg/m^3, A is the area in m^2 intercepting the wind, V is the wind speed in m/s, and C_D is the coefficient of drag. A cylinder has a coefficient of drag of 0.6, and a flat plat has a coefficient of 1.1. For a small wind turbine rotor, it is best to assume that it acts like a flat plate [2]. Remember to add the wind loading on the tower itself because it is also subjected to wind loading.

International standards recommend that the tower be designed to withstand the maximum wind loading expected for the critical design wind load [3]. This wind speed is usually 50 m/s for most cases. Most small wind turbines will indicate that they are designed to with stand 50 m/s or 120 mph. Many tower companies will use a safety factor of two in their tower designs.

The second important consideration is the dynamic interactions between rotational vibrations and natural resonance frequencies of the tower. Continuous operation at some rotational speeds of the rotor might cause the tower to oscillate to a point that causes the system to become unstable and collapse. The tower needs to have sufficient stiffness to prevent these oscillations or the rotor speed needs to be adjusted. Towers recommended by wind turbine manufacturers have usually been examined to meet both of the considerations discussed. Although there are no specific requirements for towers specified in small wind turbine standards, there are two tests in the certification process that examine the interactions of the tower and wind turbine. The duration test requires that any excessive tower vibrations be noted and that the applicant provide an evaluation of potential dynamic interactions between tower and turbine. They must show that potentially harmful dynamic interactions will be avoided [1]. The safety and function testing also notes any tower dynamic interactions with the turbine during start–stop operations.

The most common problem with selecting towers for small wind systems is that they are usually too short. Many small wind systems were originally modeled after the American water pumping windmill. These mills were placed in remote areas with few obstructions and worked well on short towers of 10 meters (33 feet). Another reason short towers worked for water pumping systems is that they reached their maximum pumping rate at 9 m/s wind speed and the rotor furled out of the wind at higher wind speeds. Many small electric-generating turbines reach their rated power at 11 m/s and continue to produce maximum power up to 18 m/s before furling or the output is regulated. Newer wind systems just need to operate in higher winds than the older water pumping designs. Most installers that the author has discussed tower heights with have agreed that small systems need to be on towers of 25 to 35 meters (80 to 120 feet). Figure 4.1 shows two wind turbines and potential problems when towers are too short. The turbine on the short tower just will not perform well because it is shielded from good winds by the trees.

Towers come in many shapes, styles, and designs. The author has chosen to divide towers into two classifications of guyed and freestanding for ease of discussing the many types. In each of these general categories, the author

Figure 4.1 Two wind turbine towers showing that taller towers are better suited to get the turbine into the full wind flow. Poor performance is often related to improper tower selection and installation.

discusses the various types of construction used and also foundations because foundations are significantly different for these two general types of towers.

GUYED TOWERS

Guyed towers are supported by a number of supporting cables connected at various points up the tower. There are usually three or four cables for each guy-connecting location. Guyed towers are usually smaller in size and are made of either tubular material or a framework of small rods and pipes that look like a lattice. Guyed towers are popular for small wind machines because they offer good strength, are easy to install, have a pleasing appearance, and are reasonably priced. Main disadvantages to using guyed towers are the larger area required for the supporting guy cables and the difficulty of maintaining the area around the guy anchors and cables. Figure 4.2 shows a small wind turbine on a guyed lattice tower that is 24 meters (80 feet) tall. Note that this tower has two guy-connecting points, one just below the tip of the rotor blades and another at about 16 meters (50 feet) or slightly above half the height. For taller towers, manufacturers will typically recommend three

Figure 4.2 A small wind turbine on an 18 meter (60 feet) lattice guyed tower with two levels of guys.

guy-connecting points. Figure 4.3 shows a small wind turbine on a tubular tower. Tubular towers may be called monopole towers, pipe towers, or other similar names. It is important to note that tubular towers should be increased in diameter to increase stiffness instead of using heavier wall thickness. A lightweight tower that is 20 cm (8 inches) in diameter is stiffer than a 10 cm (4 inch)–diameter heavy wall pipe. Most of the time a tower needs to be stiffer to reduce the potential for dynamic vibrations. Note that this tower has one level of guys instead of two like the lattice tower.

Foundations for Guyed Towers

The base foundation of a guyed tower is designed to support the weight of the tower and any downward force caused by the loading of tension in the guy cables. The thrust or overturning loads on the guyed towers are carried by the guys and the anchors. The guy anchors have to be large enough to support the overturning thrust loads. Because of this configuration, the guy

Figure 4.3 Small wind turbine on a 24 meter (80 feet) tubular guyed tower with one level of guys [7].

anchors may sometimes be as large or larger than the tower base. Figure 4.4 shows the cross section of a base foundation for a guyed tower. The design usually includes a larger bottom section than the actual foundation to increase the resistance to the downward force. Oftentimes if soil proprieties allow, a foundation drill is used to drill a round hole with a flared bottom and the complete hole is filled with concrete. This reduces the forming required and offers better contact between the soil and the concrete. Figure 4.5 is a center base ready for pouring and a completed center base with appropriate grounding for lightning protection. Tower and wind turbine manufacturers should supply the purchaser with appropriate foundation guidelines for the tower used.

Guy anchors are an important component of any guyed tower installation. They should be installed to fully support the expected loading on the tower for any designed loading, typically for wind speeds up to 50 m/s (120 mph). Figure 4.6 shows some examples of acceptable guy foundations. Placing the concrete block at least a meter deep will almost double the uplift capacity of the concrete block. Another design that is sometimes used is to

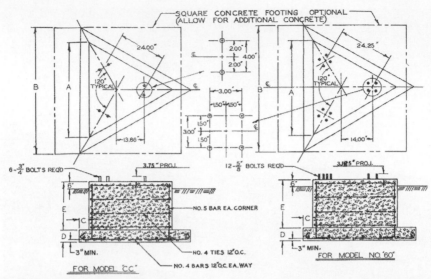

Figure 4.4 Foundation plan for the base of a guyed tower. Dimensions vary depending on height and size of tower [8].

Figure 4.5 A base foundation ready for concrete (left) and a base attached to a foundation (right). Copper wire is grounding for lightning protection.

drill a pier hole similar to what was done for the base and add a guy anchor at the specified angle with the resistance plate in the middle of the guy anchor. Alternatively, j–bolts can be inserted into the concrete pier and an adapter plate can be bolted to the top of the pier. It is important to make sure that the volume of concrete used is similar to the square anchor. Screw–in anchors have been used for small wind turbine towers but have had mixed results. It is almost impossible to determine if the screw–in is installed to meet the design loads. Soil type and moisture content of the soil greatly

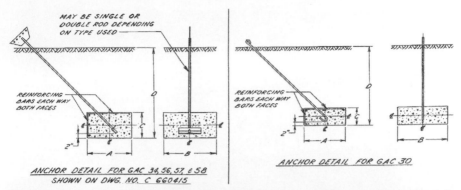

Figure 4.6 Two plans for guy anchor attachments. The anchor support on the left contains a flat plate bolted to the end of the anchor rod and the one on the right has a "J" or crook on the end. Dimensions of the concrete block depend on the height and size of the tower [8].

affect the ability to get the anchor installed to the proper depth. Several turbines have been lost due to screw-in anchors failing because of improper installation. The risks of installing a wind turbine on a tower with unknown guy anchors are just too much.

Erection of Guyed Towers

Guyed towers can be assembled on the ground and the small wind turbine attached before lifting with a crane. Care should be taken to keep rotor blades away from the lifting cable and to keep all guy wires clear of the wind turbine while lifting the tower. Many manufacturers also offer a tilt-up design that allows the tower to be lifted with the use of a winch. Carter Wind Systems [4, 5] used this technique to install several hundred 25 kW wind turbines in the 1980s using 18 and 25 meter (54 and 82 feet) towers (Figure 4.7). An essential component of a tilt-up system is the "gin pole" or a fixed section to use as a lever arm to provide leverage to lift the tower weight. Tilt-up systems have been used for both lattice and monopole towers, but is typically used for monopole towers. Lattice towers are not designed to withstand side-lifting loads as well as monopoles. A smaller guyed lattice tower can also be assembled by adding a tower section one at a time using a clamp-on lifting pulley. The tower sections usually come in 3-meter (10-feet) sections that are lifted easily with a pulley and rope and slipped over the existing sections. Guy wires are added as the height is increased to support the tower. A small wind turbine can be lifted into place at the top of the tower with the same assembly device.

Figure 4.7 Guyed towers can be adapted for erection using a tilt-up system. A stiff arm or gin pole is used as a lever to lift the tower [7].

FREESTANDING TOWERS

As the name implies, freestanding towers are not supported by cables or anything but the tower itself. A major advantage of a freestanding tower is that it has a small footprint and requires little space outside of the tower base and foundation. This type of tower is popular in urban and small space locations. However, others prefer the freestanding tower because they do not have to worry about the guy cables used in guyed towers. Figure 4.8 is a photo of a 30 meter (100 feet) freestanding tower located near a suburban convenience store and car park. For wind turbines in the 25 to 100 kW size, almost all of them use a freestanding tower because the tower loading requires large cables that become cost prohibitive. Freestanding towers also come in lattice and monopole designs. Most of the lattice towers used for small machines have three legs constructed of pipes with bolted angle framing to form the lattice design. Very few freestanding lattice towers are welded in sections like the small guyed lattice towers. Freestanding lattice towers are improved versions of the old steel windmill towers and early towers built by Jacobs, Wind Charger, and others for battery charging in the 1920s and 1930s.

A monopole freestanding tower is shown in Figure 4.9. These towers are typically tapered for added strength and are built of heavy gauge metal for added strength. The tower has to be strong enough to prevent buckling and

Figure 4.8 A small wind turbine on a 30 meter (100 feet) freestanding lattice tower.

be stiff enough to avoid dynamic vibrations. Towers taller than 20 meters (66 feet) normally come in sections that must be assembled at the wind turbine site. Connection techniques vary from a flange-to-flange connection to an insert slip-type connection. Access to the tower top of monopole towers is usually by crane or man-lift. Larger monopole towers used on turbine sizes above 50 kW may be large enough to have a ladder mounted on the inside of the tower. Many of the earlier wind turbines manufactured in Denmark used this type of monopole tower on their 65 and 100 kW machines. Figure 4.10 shows a 100 kW turbine mounted on a monopole tower that has a ladder mounted inside the tower.

Foundations for Freestanding Towers

Freestanding towers have only one foundation and it must be sufficient to support the downward load of the tower and turbine plus the overturning loads on the tower and turbine. This means that this foundation is much

Figure 4.9 A small wind turbine on a 16 meter (54 feet) monopole tower.

larger than any of the foundations for guyed towers. Foundations for lattice freestanding towers usually use a pier-type system with a pier under each leg. Figure 4.11 shows a foundation ready for the concrete. This foundation was for a 50 kW wind turbine on a 25 meter (80 feet) tower. The piers were 1 meter (3 feet) in diameter and 5 meters (16 feet) deep with a belled bottom. The top cap that connected the three piers was 20 cm (8 inches) thick and was used to support control equipment and to prevent weeds from growing beneath the wind turbine. This foundation is typical for three-legged freestanding lattice towers.

Foundations for monopole freestanding towers usually consist of a single block of concrete that can be a square block, cylindrical in shape, or an inverted "T" shape. The square block design is used mainly for towers with small turbines, less than 10 kW, and tower height less than 25 meters (80 feet). The block can be square, rectangular, or even round. Mounting bolts are usually set in the middle of the block (Figure 4.12). The weight of

Figure 4.10 A turbine on a monopole tower that offers a ladder inside the tower. This foundation is mounted on piers (used in Alaska).

Figure 4.11 Foundation for a freestanding lattice tower with three legs before being filled with concrete.

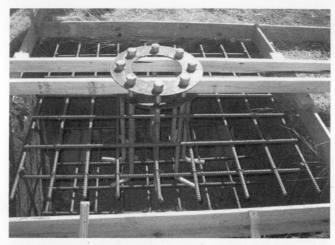

Figure 4.12 A square block foundation for a freestanding tubular tower with mounting bolts. Used for wind turbines less than 10 kW and less than 25 meters (80 feet) in height [9].

the block must be sufficient to prevent overturning of the tower. Midsize turbines (10 to 100 kW) use a foundation that is shaped like a cylinder or a large pier (Figure 4.13). The cylinder is usually drilled to a depth of 5 to 7 meters (15 to 25 feet) and diameters up to 3 meters (10 feet). Again the size of the cylinder is dependent on the size of turbine and tower used. Mounting bolts are typically in a circular pattern for attaching the bottom flange (Figure 4.14). A version of the cylinder design is a tube where the center of the cylinder is backfilled with soil instead of filled with concrete. A solid concrete cap is placed over the entire cylinder. The inverted "T" design looks like an inverted mushroom and is used mainly on megawatt size machines, but the design might be useful in some unusual soil conditions. The design consists of a large base that is 1 to 1.5 meters (3 to 5 feet) thick with a cylinder in the center that is approximately 2 meters high. The cylinder contains the bolts for supporting the tower, and the large base is covered with soil so that only the cylinder is exposed. The base should be four to five times the diameter of the cylinder to provide a large base. This design provides overturning strength without digging deep into the soil. Locations with rocky soil where excavation is difficult or soils with high water tables are potential sites for this type of foundation. Engineers have developed several special foundations for locations that have unique soil conditions that require special treatment. In Figure 4.10, the foundation is actually several piers that support a large concrete base. This foundation

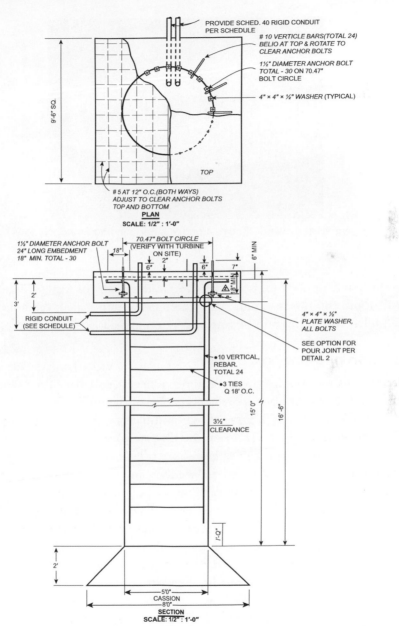

Figure 4.13 A cylinder foundation used for wind turbines in the 10 to 100 kW size. Depth and diameter are adjusted depending on the size of turbine and tower height.

Figure 4.14 A cylinder foundation for a freestanding monopole tower.

system was designed for installing towers in the tundra of Alaska. Holes were drilled into the frozen tundra; after the piers were installed, the holes were backfilled with a mixture of sand and water and allowed to refreeze.

Erection of Monopole Towers

Erection of monopole towers requires a crane or similar lifting device to raise the tower into place. For small machines, the turbine is often installed on the tower while it is lying down and the complete unit is lifted into place. Care should be taken to ensure that the rotor blades do not come in contact with the lifting cable. For some taller towers and where space is limited, towers are assembled in sections and the sections are attached one at a time. Lattice towers can usually be assembled in 6 or 12 meter (20 or 40 feet) sections and then erected one section at a time. When this is done, personnel must be able to access the connection location by either climbing the tower or using another man–lift machine. Once the tower is attached to the foundation, it must be leveled and secured in place. All tower manufacturers have specifications for attaching the tower to the foundation. After the tower is in place, the wind turbine can be attached to the top and all electrical wiring completed. A tilt–up system is not normally used with monopole towers because there are no guy wires to keep the tower from swinging back and forth during the tilt–up process. Tag line ropes are sometimes used on towers less than 20 m tall to keep the tower in position during the tilt–up procedure. This is not recommended unless done by experienced personnel.

TOWERS FOR VERTICAL AXIS AND OTHER TURBINES

Several other wind turbine designs are being marketed that require special towers that do not exactly match the descriptions and load requirements of guyed and freestanding towers. The Darrieus vertical axis design is the only vertical axis design that has been manufactured on a large scale, with several hundred being erected in the mid- to late 1980s [5, 6]. This design used a three-legged lattice tower with supporting guys from the top cap mounted above the turning rotor (Figure 4.15). Design loads for the foundations would be similar to loads for a guyed lattice tower where the base carries the downward load and the guy anchors carry the overturning loads. Towers for straight-sided vertical axis turbines (gyromills) often use a lattice design that is freestanding (Figure 4.16). Some of the newer designs

Figure 4.15 A vertical axis wind turbine of the Darrieus design showing a short tower and long guy wires supporting the top of a wind turbine.

Figure 4.16 A vertical axis wind turbine of the gyromill design on a freestanding lattice tower. Turbine is less than 6 meters (20 feet) tall.

with curved blades use a monopole tower design. Most of these turbines are less than 4 meters (13 feet) in height and 2.5 meters (8 feet) in diameter and weigh less than 200 kg (450 lb). These small turbines are mounted on relatively short towers of less than 15 meters (50 feet). It is important to purchase a tower with these wind turbines because they have some unusual dynamic characteristics that make it difficult to match a tower to the wind turbine (Figure 4.17). Purchasing the system as a package with turbine and tower places the responsibility on the manufacturer to ensure that the tower will not experience unacceptable behavior.

One popular use of small wind machines is for charging batteries to power street lights. Figure 4.18 shows how a street light tower was adapted

Figure 4.17 Several novel designed wind turbines have chosen freestanding monopole-type towers for their turbines [7].

Figure 4.18 Combine small wind turbines with solar energy to power outdoor and safety lighting in remote areas.

to add a small wind turbine and a solar panel for powering a safety light. Several companies offer a complete package system where a wind turbine is combined with a lighting system. This is another example of specially designed towers to fit the needs and desires of customers.

Figure 4.19 Roof-mounted wind turbine. Performance will be poor and not recommended [10].

Roof-Mounted Towers

Placing wind turbine towers on the roofs of houses and buildings is never a good idea. Towers transmit the vibratory motions of the wind turbine to the earth and vibratory motions are transmitted through any structure in the path. Oftentimes small wind turbines that operate at variable speeds will be operating at a speed that amplifies the vibratory motions. This makes the building vibrate, causing an irritating noise, or begins to create cracks in walls or other components. Vibration is just one issue concerning roof-mounted wind turbines. The performance is always lower because of interference from the building, reducing the wind speed available to the wind turbine. Also, trees grow and will block a roof-mounted wind turbine after a few years. Figure 4.19 is an example of poor planning when placing a turbine on a roof. Much of the incoming energy to the turbine is blocked. Sadly, this is more typical than the exception for roof-mounted turbines, regardless of the size.

REFERENCES

[1] American Wind Energy Association. AWEA Small Wind Turbine Performance and Safety Standard. AWEA 9.1–2009. Washington, DC: AWEA; 2009.

[2] Gipe P. Wind Power, Renewable Energy for Home, Farm, and Business. White River Junction, VT: Chelsea Green Publishing; 2004.

[3] International Electrotechnical Commission. International Standard. IEC 61400–2, Wind Turbines–Part 2: Design requirements for small wind turbines 2006.

[4] Southern California Edison Co. San Gorgonio Wind Parks; 1986.

[5] Southern California Edison Co. Tehachapi Wind Parks; 1986.

[6] Pacific Gas and Electric Co. Altamount Pass Wind Parks; 1988.

[7] Alternative Energy Institute, West Texas A&M University, Canyon, TX.

[8] Rohn Products LLC, Peoria, IL.

[9] Photo gallery, Usethewind.net.

[10] The Des Moines Register, June 19, 2009.

CHAPTER FIVE

Machine Selection
Matching the Machine to the Site and Power Needs

Selecting a wind machine to perform the work needed to meet your requirements is an iterative process that may take several attempts before deciding on a final machine. There are many things to consider, such as the size turbine to produce the required or wanted power, one that is aesthetically correct for your location, one that has a good reliability rating, one that has a local dealer to perform periodic maintenance, and finally the cost. One must remember that because a wind machine is made of many parts, there are many different choices where each is designed to meet the various needs and desires of the customer. You also have to remember that a wind machine needs a good energy source so having sufficient wind to power the unit is essential. Many good machines have not been successful simply because they were not placed in the correct location. However, this chapter considers the machine characteristics needed to make the machine work for the selected application.

The selection process should consider four important areas; however, many individuals only consider one or two of them. Performance (1) is often considered the most important area and is usually the only one considered, followed closely by the attractiveness (2) or the aesthetics of the machine. Noise (3) and reliability (4) are equally important, but are normally not considered as important as performance. Standards for small wind machines include discussions of three of these four elements that the author considers important in the selection process. Performance, structural integrity (reliability), and noise are included in the standards, and any machine certified to meet one of these standards will have undergone testing to ensure that it meets the specified requirements. International Electrotechnical Commission Standards [1] focus mainly on structural designs, reliability, and performance. Both the British Wind Energy Association's Standard [2] and the American Wind Energy Association's Standard [3] include performance, reliability, safety, and acoustic (noise). Attractiveness is dependent on each person's perception of what a wind machine should look like; therefore, it is not included in anyone's standard.

SELECTION USING PERFORMANCE INFORMATION

This is the time when manufacturer's performance data and/or performance data from certification testing are used to determine the size of wind turbine needed. Performance data are typically reported in one of two formats: rated power and as a power curve. Many manufacturers use a rated power number to describe the performance of their machines. Rated power is the instantaneous power at a given wind speed. Standards recommend that rated power be reported for a wind speed of 11 m/s. This number gives no information about the output of the machine at wind speeds above or below 11 m/s. The author recommends not using rated power because most small wind machines will not operate near an 11 m/s wind speed for much of the time. You also need to know the efficiency of the wind turbine, which is usually not available to estimate annual energy production. It is much better to use the power curve because it gives more information about the overall performance of the machine. The shape and slope of the power curve are needed to estimate the energy production from any wind speed distribution. Many manufacturers will include a power curve with their product literature and one is also included in certification reports. Table 5.1 shows an example of annual energy being calculated using a power curve and a wind speed distribution. This method can be used to calculate by year, by month, or by any other time period desired. Annual energy comparisons are usually sufficient for most cases unless (1) there are high seasonal loads and (2) monthly or seasonal time periods need to be considered.

Now that certification reports are becoming available for several small machines, the annual energy production is determined for several different annual average wind speeds. Table 5.2 shows a typical annual energy report from the Small Wind Certification Council [4]. It contains annual energy production at eight different hub heights and annual average wind speeds, all corrected to sea level air density. Data in both Table 5.1 and 5.2 are for the same wind turbine; these data can be compared to show how the calculated annual energy using a wind speed distribution and the certification table agree. The average wind speed for data in Table 5.1 is 7.3 m/s and, by using a linear interpolation between the energy for 7 and 8 m/s in Table 5.2, yields an estimated annual energy of 33,865 kWh as compared to

Table 5.1 Energy calculations using wind speed histogram and power curve[a]

Wind speed bin (m/s)	Hours	Power (watts)	Energy (kWh)
1	189	0	0
2	241	0	0
3	487	102	50
4	744	399	297
5	987	848	837
6	1,160	1,510	1,752
7	1,199	2,403	2,881
8	995	3,602	3,584
9	796	5,071	4,037
10	602	6,856	4,127
11	417	8,863	3,696
12	304	10,885	3,309
13	215	12,019	2,584
14	153	12,395	1,896
15	103	12,495	1,287
16	68	12,546	853
17	45	12,503	563
18	27	12,442	336
19	14	12,208	171
20	14	11,989	168
	8760		32,427

[a]Average wind speed was 7.3 m/s at height of 25 m. Power curve was for a Bergey Excel 10 wind turbine.

Table 5.2 Tabulated annual energy (AEP) for a Bergey Excel 10 wind turbine[a]

Hub height annual average wind speed (m/s)	AEP measured (kWh)	Standard uncertainty in AEP (kWh)	Standard uncertainty in AEP (%)	AEP extrapolated (kWh)
4	7,135	503	7.05	7,135
5	13,842	884	6.39	13,842
6	22,300	1,281	5.74	22,300
7	31,342	1,604	5.12	31,342
8	39,755	1,824	4.59	39,755
9	46,652	1,944	4.17	46,652
10	51,626	1,982	3.84	51,626
11	54,685	1,961	3.59	54,685

[a]Data were corrected to a sea-level density of 1.225 kg/m^3.
Source: Small Wind Certification Report [4].

32,427 kWh for the total from Table 5.1. As a result, it is fairly easy to obtain performance information about a particular wind turbine, especially now that certification reports are available for several machines. Data are reported as a rated power, a power (performance) curve, or an annual energy number for a given wind speed.

MATCHING ANNUAL ENERGY CONSUMED TO ENERGY PRODUCED

Using the annual energy required for your application, which was determined using the methods described in Chapter 2, and the annual energy production of the wind turbine, it is a matter of matching the numbers—if you know your average wind speed at hub height. It may take several attempts to find a wind turbine in the right size range to match up to your energy requirements. However, you also need to consider the type of interconnection contract you have or are going to have with your local electric provider. Table 5.3 lists three different types of interconnection agreements. These are generalizations of the many agreements being used, but should provide guidance on how to adjust the ratio of energy produced to energy consumed for the various types of agreements. The basic agreement is an agreement that pays a fraction of the cost of electricity for any excess that is fed back into the utility system. This price is often called the "avoided cost," but is usually much less than the fuel cost adjustment. With this type of contract, the wind turbine production should be much smaller than the consumption to avoid "giving away" the excess energy generated. The second type of agreement is a net-metering agreement with a monthly "true-up." With this agreement, again the wind production should be less than the consumption because the seasonal variations in both wind energy production and electrical use will not match, causing overproduction in some months and

Table 5.3 Suggested loading fractions for three utility interconnection agreements

Type of agreement	Basic interconnect agreement	Net metering agreement with monthly true-up	Net metering agreement with yearly true-up
Fraction of annual energy consumption	One-half	Three-quarters	Full use
Percentage of annual energy consumption	40–60%	60–80%	80–100%

underproduction in other months. Extra wind energy in the spring does not count toward high electrical use in the summer. Finally, the agreement with an annual "true-up" is the agreement where you want to closely match wind turbine production with annual energy consumption. For most years, you should be able to balance the energy consumed with the energy produced. Generally, you cannot make the wind system economically feasible by selling power at a reduced rate to the utility. You must always receive full retail price

Table 5.4 Monthly electrical energy purchased, generated, and consumed by all-electric home with small wind turbine

Month	Meter		Turbine produced (kWh)	Energy	
	Purchased (kWh)	Sold (kWh)		Gross used (kWh)	Net used (kWh)
Jan-09	5,398	43	313	5,668	5,355
Feb-09	3,316	100	366	3,582	3,216
Mar-09	2,341	141	437	2,637	2,200
Apr-09	2,078	128	464	2,414	1,950
May-09	1,176	50	271	1,397	1,126
Jun-09	1,335	25	217	1,527	1,310
Jul-09	2,001	16	222	2,207	1,985
Aug-09	1,547	68	314	1,793	1,479
Sep-09	1,233	45	226	1,414	1,188
Oct-09	1,455	76	321	1,700	1,379
Nov-09	2,264	71	311	2,504	2,193
Dec-09	6,741	29	304	7,016	6,712
Year total	30,885	792	3766	33,859	30,093
		21%	11%		
Jan-10	6,019	28	238	6,229	5,991
Feb-10	5,129	12	144	5,261	5,117
Mar-10	2,721	56	467	3,132	2,665
Apr-10	2,051	237	596	2,410	1,814
May-10	973	136	427	1,264	837
Jun-10	1,244	45	367	1,566	1,199
Jul-10	691	69	276	898	622
Aug-10	2,251	7	224	2,468	2,244
Sep-10	1,159	44	264	1,379	1,115
Oct-10	1,138	51	244	1,331	1,087
Nov-10	2,217	86	344	2,475	2,131
Dec-10	3,232	68	365	3,529	3,164
Year total	28,825	839	3,955	31,941	27,986
		21%	12%		

Source: Data provided by Byron Neal, Canyon, TX, 2012.

for the energy you produce. This is accomplished by selecting a machine that is smaller than the load for most utility connection agreements. The type of agreement impacts what fraction of the load you want to displace to maximize your return.

Table 5.4 contains two years of actual energy production from a small wind turbine, energy purchased for a rural home, excess energy sold back to the utility, total energy consumed in the home, and net energy used for net billing. These data are from an all-electric home with a water well, barns, and livestock watering for less than 10 animals. The home also has seven occupants and is located in the southern Great Plains of the United States. These data clearly show the issue of using annual average data or monthly average data to match to a system where energy is measured instantaneously or in 15-minute increments. Even in months of high energy consumption there are times that the wind turbine will be producing more than is consumed. Note in December 2009 and January 2010, both high energy use months, there was still 29 and 28 kWh of excess wind power. For this example, 20% of the wind power was fed back to the utility system for others to consume. This home had an electrical contract with net metering using a monthly "true-up." The home owner actually paid for the amount in the net energy used column. However, if he had a basic electrical interconnection contract, he would receive a fraction of the retail cost and the value of the energy sold would be minimal. This is why the author recommends not oversizing your machine depending on the type of interconnection contract that is available. The machine in Table 5.4 is undersized for the load, but it still provided a good return for the owner's investment.

SIZING A MACHINE FOR OFF-GRID APPLICATIONS

Matching a wind machine to a load for off-grid applications is much different than matching one to work interactively with the electric grid. If the wind turbine does not provide sufficient torque to start the motor or load, then the wind turbine will stall and no work will be done. Also, if the wind turbine does not maintain sufficient output, the load will overpower and stop the wind turbine. The easiest way to remember this issue is to think of the load or motor as a brake on the system. Clark and Vick [5] developed guidelines for matching electric water pumps to various sizes of wind turbines. These were pumps used for livestock watering, domestic water use, and small-scale irrigation systems when no electric grid was available. They recommend that the wind turbine have an approximate rating of twice the rated size of the pump motor. This larger wind turbine was needed to provide sufficient electrical current to start the loaded pump motor. However, if the system is a battery–charging system whereby batteries are used to store excess energy when the wind is providing excess energy, then the wind turbine can be more nearly matched to the electrical load. Off-grid systems typically require that the wind turbine be larger than the rated load, whereas on-grid systems need to be smaller than the rated load.

MACHINE NOISE, RELIABILITY, AND SAFETY INFORMATION

Once you find a machine that matches your energy needs, then it is time to search for information on the other selection parameters. If a certification report is available for the machine you have selected based on size, then you can easily check the information on noise, reliability, safety, and function testing. Figure 5.1 provides a copy of a published certification report for the Bergey Excel 10 [4]. Acoustic data are given in sections 7 and 8 shown in Figure 5.1 and show that the noise level was measured at 42.9 decibels. The report shown in Figure 5.1 also contains a chart of actual data with the total noise level with the turbine operating and the noise level with the turbine off (background noise). These data can be compared between machines and you can choose one that is less noisy if this is a critical issue for you. Several manufacturers provide some information in their product literature about noise from their machines. However, noise is a relative term and each of us has a different perspective of what is noisy.

Another consideration is the reliability of the wind machine. A machine that has not been fully tested is subject to having operational problems. In order for a small wind machine to be certified, it has to complete a 2,000–hour duration test, which includes several hours of operation at each wind speed in ranges from 5 to 20 m/s [4]. All problems or repairs during the duration test have to be reported in the certification report. This information is reported in section 9 (Figure 5.1). If no certification report is available, inquire about the number of machines sold and how long they have been in operation. Any manufacturer who has made less than 20 machines is probably still in the developmental stage and their machines are still being modified with each new model. An ongoing problem with new technologies such as small wind machines is that

SWCC Summary Report

Manufacturer: **Bergey Windpower Company**

Wind Turbine: **Excel 10** (240 VAC, 1-phase, 60 Hz)

Certification Number: **SWCC-10-12**

The above-identified Small Wind Turbine is certified by the Small Wind Certification Council to be in conformance with the AWEA *Small Wind Turbine Performance and Safety Standard* (AWEA Standard 9.1 – 2009).

For the SWCC Certificate visit: www.smallwindcertification.org

CERTIFIED
SMALL WIND TURBINE
SWCC-10-12

1. Introduction

This report summarizes the results of testing and certification of the Bergey Excel 10 in accordance with AWEA Standard 9.1-2009. The Excel 10 is a 3-blade, upwind, horizontal axis wind turbine with a swept area of 38.5 m^2. The tested configuration utilized a Powersync II inverter and a Bergey 30 m (100 ft) guyed-lattice tower. Field tests were conducted at the USDA/ARS facility in Bushland, Texas from June 24, 2010 to March 18, 2011.

2. Turbine Ratings

AWEA Rated Annual Energy @ 5 m/s	**13,800**	kWh
AWEA Rated Sound Level	**42.9**	dB(A)
AWEA Rated Power @ 11 m/s	**8.9**	kW

3. Tabulated Annual Energy Production (AEP)

Corrected to a sea level air density of 1.225 kg/m^3

Hub Height Annual Average Wind Speed (m/s)	AEP Measured (kWh)	Standard Uncertainty in AEP (kWh)	Standard Uncertainty in AEP (%)	AEP Extrapolated (kWh)
4	7,135	503	7.05	7,135
5	13,842	884	6.39	13,842
6	22,300	1,281	5.74	22,300
7	31,342	1,604	5.12	31,342
8	39,755	1,824	4.59	39,755
9	46,652	1,944	4.17	46,652
10	51,626	1,982	3.84	51,626
11	54,685	1,961	3.59	54,685

SWCC-10-12   1

Figure 5.1 Small Wind Certification Council Summary Report SWCC-10-12 for Bergey Excel 10.

manufacturers make 5 or 10 machines at a time and then, after operating them for a few months, make design changes and build another 5 or 10 machines that are slightly different. This creates a problem for long-term operation because there is no supply of repair parts because so few

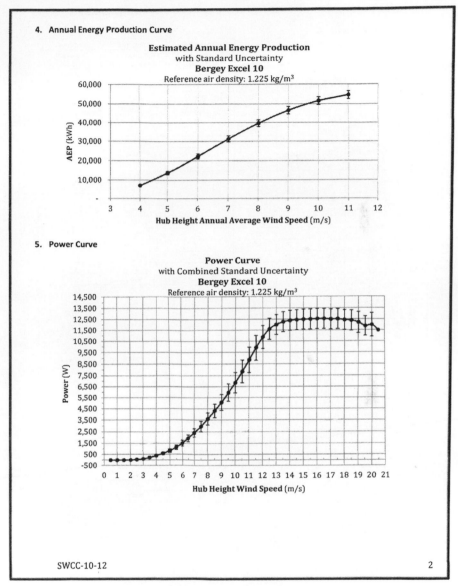

4. Annual Energy Production Curve

Estimated Annual Energy Production
with Standard Uncertainty
Bergey Excel 10
Reference air density: 1.225 kg/m³

(Y-axis: AEP (kWh); X-axis: Hub Height Annual Average Wind Speed (m/s))

5. Power Curve

Power Curve
with Combined Standard Uncertainty
Bergey Excel 10
Reference air density: 1.225 kg/m³

(Y-axis: Power (W); X-axis: Hub Height Wind Speed (m/s))

SWCC-10-12   2

Figure 5.1 (*Continued*)

machines are alike that there are no incentives to make or build replacement components.

One aspect of the reliability issue that is often overlooked is the availability of someone to provide routine maintenance and repairs. Wind turbines are complex and require knowledgeable personnel to provide service and repairs.

6. Tabulated Power Curve

Bin No.	Hub Height Wind Speed	Power Output	Cp	1-minute samples	Category A Standard Uncertainty, Si	Category B Standard Uncertainty, Ui	Combined Standard Uncertainty, Ci
	Corrected to a sea level air density of 1.225 kg/m³						
	m/s	Watts			Watts	Watts	Watts
1	0.5	-12		158			
2	1.0	-12		224	0.1	0.9	0.9
3	1.5	-11		309	0.3	0.9	1.0
4	2.0	0		391	0.9	2.9	3.0
5	2.5	39	0.11	375	2.1	10.9	11.1
6	3.0	102	0.16	661	3.0	20.2	20.4
7	3.5	229	0.23	818	3.4	43.8	43.9
8	4.0	399	0.26	1060	3.2	65.4	65.4
9	4.5	596	0.28	1213	3.0	84.5	84.6
10	5.0	848	0.29	1235	3.7	116.9	117.0
11	5.5	1,151	0.29	1279	4.7	152.6	152.6
12	6.0	1,510	0.30	1250	5.4	195.2	195.3
13	6.5	1,938	0.30	1401	6.0	248.5	248.6
14	7.0	2,403	0.30	1355	7.1	293.3	293.4
15	7.5	2,949	0.30	1014	9.9	362.8	362.9
16	8.0	3,602	0.30	885	12.7	452.4	452.6
17	8.5	4,306	0.30	687	16.8	523.1	523.3
18	9.0	5,071	0.30	736	18.0	604.1	604.4
19	9.5	5,960	0.29	668	19.7	725.9	726.1
20	10.0	6,856	0.29	707	21.4	790.8	791.0
21	10.5	7,849	0.29	650	26.2	912.1	912.5
22	11.0	8,863	0.28	599	28.0	994.0	994.4
23	11.5	9,928	0.28	635	24.3	1098.6	1098.9
24	12.0	10,885	0.27	606	24.8	1105.8	1106.1
25	12.5	11,619	0.25	504	21.7	1044.8	1045.0
26	13.0	12,019	0.23	432	15.0	968.6	968.7
27	13.5	12,276	0.21	337	13.3	906.1	906.2
28	14.0	12,395	0.19	333	7.4	906.0	906.1
29	14.5	12,449	0.17	292	7.2	904.5	904.6
30	15.0	12,495	0.16	279	3.3	907.5	907.5
31	15.5	12,508	0.14	231	10.3	907.4	907.4
32	16.0	12,546	0.13	187	5.4	911.0	911.0
33	16.5	12,555	0.12	165	8.5	910.7	910.8
34	17.0	12,503	0.11	125	24.4	908.8	909.1
35	17.5	12,528	0.10	138	17.8	909.2	909.4
36	18.0	12,442	0.09	98	36.2	908.2	908.9
37	18.5	12,396	0.08	94	36.8	901.0	901.7
38	19.0	12,208	0.08	57	65.2	916.2	918.5
39	19.5	11,878	0.07	39	83.4	960.0	963.6
40	20.0	11,989	0.06	18	130.0	882.0	891.5
41	20.5	11,495	0.06	15	124.6	1066.4	1073.7

SWCC-10-12   3

Figure 5.1 (*Continued*)

Will there be someone within a reasonable travel distance to provide any service that may be required? Most manufacturers recommend that their units be inspected and checked at least once a year. The author suggests checking most machines twice per year—once in the fall before winter season and again in late spring after most of the higher winds of winter and spring. The timing

7. **Tabulated Acoustic Data**

Wind Speed @ 10m Height m/s	Background Sound Pressure Level (SPL) dB(A)	Corrected Bergey Excel SPL dB(A)	* indicates delta dB between 3 & 6 dB	Bergey Excel SPL Std. Dev. dB	Corrected Sound Power dB(A)
6	38.53	42.38	*	1.37	80.57
7	39.85	44.23	*	1.52	82.42
8	41.36	46.71		1.91	84.90
9	43.32	49.25		1.95	87.44
10	44.91	51.99		1.81	90.18

8. **Graphical Acoustic Data**

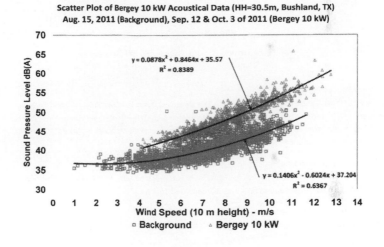

Scatter Plot of Bergey 10 kW Acoustical Data (HH=30.5m, Bushland, TX)
Aug. 15, 2011 (Background), Sep. 12 & Oct. 3 of 2011 (Bergey 10 kW)

$y = 0.0878x^2 + 0.8464x + 35.57$
$R^2 = 0.8389$

$y = 0.1406x^2 - 0.6024x + 37.204$
$R^2 = 0.6367$

Wind Speed (10 m height) - m/s

□ Background   △ Bergey 10 kW

Figure 5.1 (*Continued*)

should be adjusted depending on the type of wind resource at the wind turbine site. A third consideration that is part of reliability is the manufacturer's warranty. Is there a manufacturer's warranty provided and what is covered by the warranty? What is the length of the warranty? Many manufacturers offer a 5-year warranty of materials and workmanship.

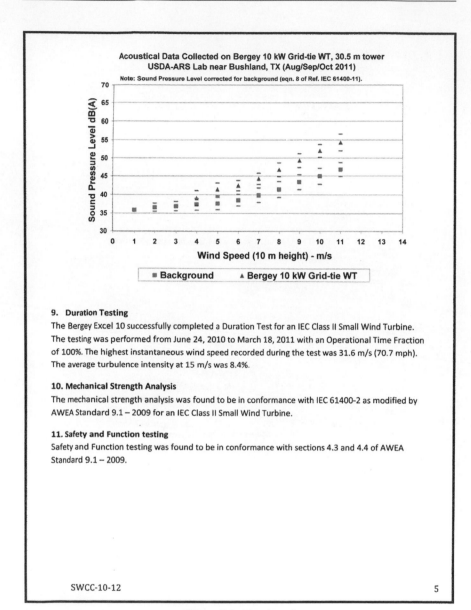

Acoustical Data Collected on Bergey 10 kW Grid-tie WT, 30.5 m tower
USDA-ARS Lab near Bushland, TX (Aug/Sep/Oct 2011)

Note: Sound Pressure Level corrected for background (eqn. 8 of Ref. IEC 61400-11).

Wind Speed (10 m height) - m/s

■ Background   ▲ Bergey 10 kW Grid-tie WT

9. Duration Testing

The Bergey Excel 10 successfully completed a Duration Test for an IEC Class II Small Wind Turbine. The testing was performed from June 24, 2010 to March 18, 2011 with an Operational Time Fraction of 100%. The highest instantaneous wind speed recorded during the test was 31.6 m/s (70.7 mph). The average turbulence intensity at 15 m/s was 8.4%.

10. Mechanical Strength Analysis

The mechanical strength analysis was found to be in conformance with IEC 61400-2 as modified by AWEA Standard 9.1 – 2009 for an IEC Class II Small Wind Turbine.

11. Safety and Function testing

Safety and Function testing was found to be in conformance with sections 4.3 and 4.4 of AWEA Standard 9.1 – 2009.

SWCC-10-12   5

Figure 5.1 (*Continued*)

Closely related to reliability is the safety and function testing required for certification. These are tests to see that the control functions that control the wind turbine in extreme winds, loss of external electricity to the controller, or other emergency situations all perform as designed to protect the wind turbine when those conditions occur. This provides assurance that the brakes

12. Manufacturer Tower Design Requirements

BASIC TOWER REQUIREMENTS for the BWC EXCEL WIND TURBINE

Customer supplied towers for the BWC EXCEL should meet the following requirements:

Tower Height:	60 ft (18 m) minimum, 80 ft (24 m) or higher recommended
Design Wind Speed:	120 mph (54 m/s)
Turbine Weight:	1200 lb (545 kg)
Turbine Thrust Load:	2400 lb (1090 kg) @ any wind >= 40 mph (18 m/s)
Blade Clearance:	The top 12 ft (3.5 m) of the tower must not extend beyond an 18 inch (0.46 m) radius from the tower centerline.
Tower Plumb Tolerance:	Up to 0.25° tolerance from plumb allowed.
Tower Stiffness:	Tilt at the top of the tower should be no more than 2.0° for consistent furling. Deflection of monopole towers at 50 mph should be no more than 1.0% of tower height; at 120 mph no more than 2.5% of tower height. (For a 120 ft tower this would be 14.4 in and 36.0 in, respectively.) Overly flexible towers can cause vibration and/or fatigue problems. A civil engineer should approve the tower.

Blade Frequency:

First Flap Frequency for 10 kw (Not Rotating)			Blade Length	
Tested: 8/4/2011				
Ferrite	3.012	Hz	128	in.
Neo	2.703	Hz	134	in.

Turbine Mounting:

- Provisions shall be made for mounting a furling winch, strain relief for tower wiring, tower climbing, anti-fall equipment and access holes where appropriate.
- The top of the tower shall be designed to allow the connection of the power cable and furling cable to the turbine via the two 2.3" diameter holes in the turbine's tower adapter plate.
- A connection shall be made between the turbine furling cable and the tower furling winch by using a tower furling cable assembly (11508-x), a 3/16" stainless steel thimble (HM3003) and two 3/16" stainless steel malleable clips (HM3002-B).
  - o Furling cable, thimble, and clips must be purchased separately.
- Tower connection shall be made using either nine 5/8" bolts or six 3/4" bolts using the pattern illustrated below:

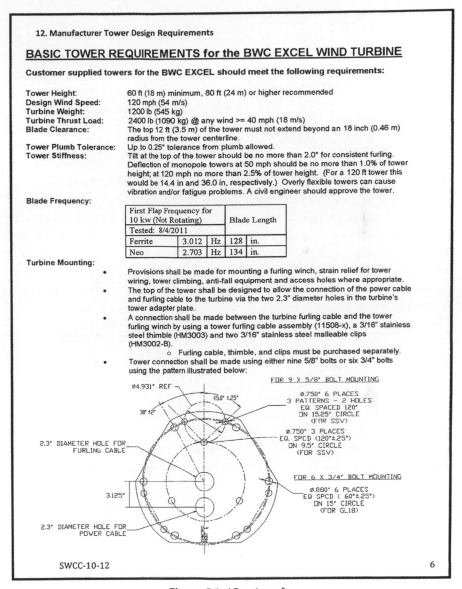

Figure 5.1 (*Continued*)

work as designed and the start–up system works properly. All of the wind turbine standards address these control functions, and there needs to be some assurance that the turbine selected will perform these functions, if needed. Proper operation of these safety functions protects not only the wind turbine and people, but the adjoining property.

AESTHETICS

There have been many attempts to make wind machines more pleasing to the eye; however, most of these designs have not been satisfactory either because of operational efficiencies or because of structure integrity. If lawn art is what you want, then don't waste your money trying to include a wind machine in the art. A well-designed wind machine with smooth lines can be easy on the eye without making it with butterfly wings or flower petals for rotor blades. The laws of physics just do not allow many deviations from high-performance airfoils. There are reasons why wind turbines look similar—just like airplanes all look similar. Another component of the wind system that can make installation look good is the right selection of a tower. Use of a cylindrical tower in place of a lattice tower might make the installation look much more appealing. A freestanding tower is sometimes more appealing than a guyed tower because of the numerous guy wires associated with guyed towers. All of these factors need to be considered if aesthetics are an important consideration in selecting a wind machine.

COSTS

Before making any type of investment in energy-generating equipment, there is a cost consideration to determine if the investment will provide the return you desire. Everyone should have a goal of what they plan to accomplish with the purchase of a small wind turbine system. Justifications for purchasing a machine range from wanting the latest gadget, to being the first one in your neighborhood, to actually reducing your electric bill and providing a hedge against higher electric bills in the future. Everyone should complete a good economic evaluation of the proposed installation and the long-range expectations of having a wind machine. Details on completing an economic evaluation are provided elsewhere in this book.

Comparing different wind machines will require considerable reading, study, and work before sufficient information is gathered and assembled to make comparisons between machines. This process is becoming much easier and will continue to improve as more machines complete the certification process. Certification brings most of the important data into a central location where it can be accessed easily. It will still be up to the individual purchaser or hired consultant to sort through all the information and make a sound judgment of the best machine to fit the wind resource available and the load requirements.

REFERENCES

[1] International Standard. Design Requirements for Small Wind Turbines. IEC61400–2. Geneva, Switzerland: International Electrotechnical Commission; 2006.

[2] RenewableUK. Small Wind Performance and Safety Standard. MCS006. British Wind Energy Association, www.bwea.com; 2010.

[3] American Wind Energy Association. AWEA Small Wind Turbine Performance and Safety Standard. AWEA 9.1-2009, www.awea.org; 2009.

[4] Small Wind Certification Council. Certification Report for Bergey Excel 10. www.smallwindcertification.org; 2012.

[5] Clark RN, Vick BD. Determining the proper motor size for two wind turbines used in water pumping: Wind Energy-95. ASME, SED 1995;16:65–72.

CHAPTER SIX

Permitting
Meeting the Institutional Issues for Installing a Small Wind Machine

Governments at all levels have zoning, permitting, and inspection rules that are in place to ensure the safety of structures that are being constructed. Often, these requirements seem as deterrents to getting new technology started. One must remember that these rules were often put in place after a major disaster, such as the Chicago fire in 1871 or Hurricane Katrina in 2005 in New Orleans. They are created to provide a minimum standard for construction so that people's lives are protected. Wind systems must also be constructed to ensure that the public is protected. Because small wind machines contain so many different technologies, several permits and inspections are often required, thus making the approval process seem insurmountable.

The first item to consider is the general zoning classifications that many cities, counties, and sometimes states have enacted to regulate the types of activities and construction that can take place in a given region. Some examples of general zoning are residential, commercial, agricultural, and historical. There will usually be a listing of activities that are acceptable within each zoning category, as well as activities that are prohibited. Small wind systems may be already be listed as approved in certain categories, but restricted from the other areas. Generally, small wind systems will be included in commercial and agricultural, but excluded from historical and residential. Remember, these are general zoning guidelines and can usually be modified for individual cases through a specific use permit. When requesting a specific use permit, a public hearing is almost always required. In most cases, these existing zoning rules have been in place since the mid-1980s and do not allow for tall structures, especially wind turbines. This is where you as an applicant for a permit become educator, leader, and rebel in writing a new zoning ordinance or variances to existing ordinances [1].

HEIGHT OF WIND TURBINE

A wind turbine must be located in a clear, free airstream to harvest the energy required to make it a profitable system. A tall tower enables the wind rotor to access the free unobstructed winds that produce the most energy and reduces the vibrational loading on the rotor blades. A tower needs to be a minimum of at least two times the height of any trees, buildings, or other obstructions near the wind turbine. Several manufacturers of small wind systems recommend that towers in residential areas be at least 25 meters or 80 feet tall (Figure 6.1). If tall trees are present, then tower heights of 30 to 40 meters or 100 to 120 feet are recommended [2, 3]. For detailed information on selecting tower heights, see Chapter 1.

Obtaining a permit for a tower in a residential zoned area may be difficult because many of these areas have height restrictions already included in the zoning ordinances. Even in commercial areas, a height restriction of 10 meters or 35 feet is written in many U.S. ordinances. The 35 foot restriction was chosen because of limitations in fire-fighting equipment many years ago before it was common to have ladder fire trucks. Typically, if new construction is needed, they just apply for a variance to the height restriction rather than rewriting or revising the ordinance. The small wind turbine owner is then left with the expense of seeking a variance or hiring a lawyer to rewrite the ordinance.

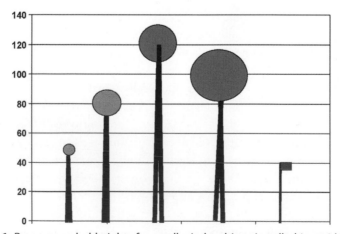

Figure 6.1 Recommended heights for small wind turbines installed in residential and commercial sites.

The need for a tall tower is probably the easiest item to justify regarding permits, but because of the height of the tower, many other concerns enter into the permitting process. Issues such as foundation requirements, tower security, aesthetics, and sound all are usually discussed in the process of discussing the need for a tall tower. Each of these will be discussed, as well as some other issues that will usually be mentioned in a public hearing.

SETBACK DISTANCE

A specific wind turbine zoning ordinance or a wind turbine construction permit should include a section describing a setback distance. Setback distance is defined as the distance from a property line, inhabited structure, utility lines, or road right of way to the wind turbine tower base. For wind turbines, the minimum setback distance should be equal to the height of the tower plus the length of one blade (Figure 6.2). Thus a wind turbine with a rotor diameter of 7 meters (22 feet) on a 30 meter tower (100 feet) would have a setback distance of 33.5 meters (111 feet).

The obvious reason for using the tower height plus half of the rotor diameter is for overtipping of the tower. Overtipping is usually caused by anchor bolt failure, guy cable failure, or tower leg failure, all of which are very rare in certified engineered towers. Certified towers are designed to withstand winds up to 55 m/s (125 mph) (IEC Standard 61400-2) [4]. Setback distances are very important when small turbines are placed on

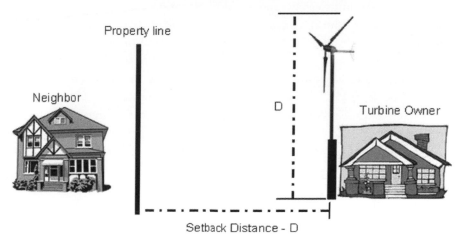

Figure 6.2 Wind turbines should be installed to meet minimum setback distances from property lines, roads, and overhead utility lines *(adapted from Stimmel [6])*.

"home–built" towers. "Home–built" towers are ones not supplied by the wind turbine manufacturer or not certified for use by the manufacturer. The setback distance has other advantages, such as providing an open uncluttered area for erection and maintenance of the wind system. It also provides a greater distance for sound to disperse.

When setback distances are used, the lot size is increased to 0.5 to 1 hectare (1–2 acres). This moves the small wind systems from most city lots to suburban areas where lot sizes are usually larger and the area is less populated. Some zoning laws may specify lot sizes or setbacks to provide a safe zone for small wind turbines placed in residential areas. Setbacks from structures may not be as important in commercial zones because structures are not normally inhabited all the time and noise may not be a critical issue.

SOUND

Many existing zoning ordinances already contain restrictions against loud noises, such as noises from truck braking and eating and drinking establishments. Because of past experience or old information circulated, writers of ordinances regulating small wind systems want to include restrictions associated with noise. Modern wind systems have reduced noise levels by using improved airfoil designs and elimination of gearboxes and other components that contributed to higher noise levels. Sound levels should be compared with background levels at selected wind speeds. Sounds from traffic, rustling leaves, and stereos can often mask or cover up the dull, low sounds coming from a small wind turbine. Also, because sound decreases with distance, the further you are from the noise source, the less the sound (Figure 6.3). The AWEA standard for small wind systems calls for a maximum sound level of 50 decibels at a distance of 60 meters (approximately 200 feet) [5]. When appropriate setbacks are observed, most small wind systems can easily meet the 50 decibels level except in rare cases when there may be loss of electrical line voltage due to a power outage and the turbine goes into an over speed condition. These events are usually rare and would also occur when the wind speed is high and the background noise is high.

Wind turbine sounds are often considered as being similar to air conditioner compressors and would not be considered as "nuisance noise" as established by many existing zoning codes. Small turbines do not produce the low thumping noise that is produced by the larger wind turbines because

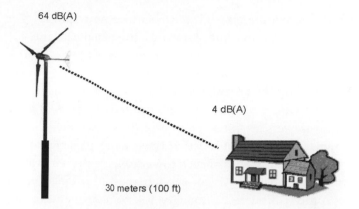

64 dB(A)

4 dB(A)

30 meters (100 ft)

Sound waves are diluted with distance

Figure 6.3 Measurements showing how sound is diluted with distance. Background noise has been subtracted from the total *(adapted from Stimmel [6])*.

the rotor blades turn much faster on small machines and do not react to the effects of the tower like the longer, bigger blades. Purchasing small wind machines that have passed certification testing will ensure that the noise levels will be at acceptable standards.

AESTHETICS

Seldom do two people see things the same way; therefore, controversy occurs about the view of the landscape when it contains wind turbines. Small wind turbines look like things of beauty, gently rotating in the sky to some, whereas others simply see them as sticks protruding up in the sky. Advocates say that small wind turbines are icons of the rural landscape, dating back to the 1700s in northwestern Europe and in the 1800s in America. Today they have emerged as a symbol of independence from dirty generated electricity. Advocates argue that as long as the turbine is installed safely and placed on private property, they should be allowed. Owners often compare a small wind turbine to a utility pole, street light, or flag pole. Communities already accept cell towers, water towers, utility poles and lines, grain silos, and radio/TV antennas as part of the landscape so why not small wind systems [6]? Thus the argument continues about aesthetics. Aesthetics are very difficult to describe in an ordinance because there are so many different viewpoints. It is often addressed in a public meeting. The owner seeking a permit to install a small wind system needs to be aware of potential negative comments and needs to prepare responses that do not "put down" the opposer, but compares the wind turbine to existing structures in the landscape (cell telephone towers, etc.)

Small wind turbine manufacturers have diligently not included large logos, advertisements, or other commercial markings on their turbines. For the most part they have been attractively painted and marked to blend into the landscape. Towers are often offered in three basic designs (monopole, lattice, and guyed monopole) to make them more attractive or blend into the landscape. The advantages of each tower type are discussed in Chapter 4, on towers and foundations. Even though they are visible, wind turbine systems are hardly considered a nuisance just because they can be seen.

FENCING FOR SECURITY

Some permits may require that the wind turbine tower be fenced to prevent unauthorized climbing of the tower. A fence seldom prevents unwanted people from climbing a tower. After reviewing thousands of tower installations, there are few instances of unauthorized climbing. A fence often imposes a false sense of security about the site and can obstruct other activities. In cases of emergencies, access to the turbine base to shut down the turbine can be slowed by the presence of a fence. Also, utilities may require that the turbine manual control be accessible to their workers for performing routine maintenance or emergency repairs to power lines. There are many other ways to prevent unauthorized climbing of a tower, such as removing the climbing rungs on the lower 3 meters (10 feet) of the tower or placing sheets of wood or metal on the lower portion of the tower to cover hand- and footholds. Using signs that give warning to high voltage or electrical shock are also used to deter potential climbers from attempting to climb towers. Fences do little more than add clutter around the base of a small wind turbine installation. Remember that someone has to trespass on private property in order to reach the tower base before they can begin to climb the tower.

ABANDONMENT

Zoning authorities may also be concerned about what happens to a wind turbine that is no longer operating. It may be stopped because of a change in ownership, malfunction of the equipment, failure of a major component, or a whole list of minor issues that may prevent the turbine from operating. The zoning authority should have a procedure in place to encourage the owner to either repair the unit or have it removed. A 6-month notification to restore the system to operating conditions or to have it removed is a reasonable time period for the owner to take action [6]. If for some reason the owner does not comply with the notification period, the wind turbine should be removed from the tower for safety reasons. A procedure similar to handling other condemned property should be followed by the zoning authority to remove the tower and other wind system components.

FOUNDATIONS

Many towns, cities, and counties now require that you have permission to perform any construction within certain boundaries. These are some of the items that should be considered when seeking a general permit to erect a small wind turbine. Once approval is granted to erect the turbine at a given location, other permits are often required for construction. Common construction permits include foundation construction and electrical wiring. Usually, both of these permits will require inspections by a local inspector at various stages of construction.

Most all zoning authorities have a permitting process for foundations, especially foundations for towers. Foundation construction is critical for the structural integrity of a tower of any kind. Foundations for small wind systems vary depending on the type of tower selected. Freestanding towers such as the monopole or lattice type will require a footing with depths up to 5 to 7 meters (15 to 25 feet) depending on the height of tower and size of wind turbine placed on the top. Guyed towers generally require a much smaller base beneath the tower and then sufficient masses at the guy anchor points to carry the tipping loads on the tower. Almost all towers purchased from a certified tower manufacturer, wind turbine manufacturer, or certified wind turbine installer will include a suggested foundation plan. The plan may be for a "worst-case soil condition" or may include three or more plans based on different soil conditions. A local registered engineer can easily determine the correct plan to be used based on the local soil conditions. The permitting authority may require a soil test at the exact location of the foundation; that information is used by the engineer to select the appropriate foundation plan.

In cases where the wind turbine owner insists on using a "home-made" tower or one that is supplied without a wind turbine certificate, care must be given to ensure that the tower foundation meets the load requirements of the wind turbine and includes a safety factor of at least two. This practice is not recommended, but some individuals will want to do their own thing. The permitting authority will still need to provide some degree of safety to the general public and neighbors; therefore, they should require an approved foundation plan prepared by a registered professional engineer. For a detailed discussion of tower types and their advantages/disadvantages, please see the Chapter 4, on towers and foundations.

ELECTRICAL PERMITS

Part of the wind turbine erection process involves connecting the wind turbine generator to the electrical load. This requires that wire be run up the tower to the generator and then underground or aboveground to the load. An electrical construction permit is usually required with periodic inspections and the requirement that the work be done under supervision of a local licensed electrician. It is important that the correct wire size be used and that all connections be made according to the prevailing electric code. In the United States, the 2012 National Electric Code for the first time included a complete section on small wind turbine installations. Section 694 includes requirements for complete installation, with detailed information on types, location, and requirements for electrical disconnects used in the system. It also includes the proper procedures for grounding to minimize damage from lightning, as well as many other recommendations to ensure a safe and functional wind installation. If the small wind system uses an inverter, only inverters that are IEEE 1547 or UL 1741 compliant should be used. Various utilities have different requirements for connecting the system to the utility grid. It is important that all the electrical wiring be done to match the wind turbine manufacturer's specifications and to meet all electrical codes.

UTILITY INTERCONNECTION PERMIT

Many small wind systems are installed with an electrical grid intertie, which means that the owner will need to secure permission to connect to the electric utility system. Small wind machines are normally connected on the customer's side of the metering system and do not require the expensive equipment that larger systems use. There are many different types of contracts depending on the individual requirements of each utility. A technical description of electrical interties is discussed in the chapter on grid integration, and the contractual processes are discussed in the following paragraphs.

Many utilities will do a simple interconnect agreement with wind turbines less than 25 kW where the owner may connect to the utility line through appropriate disconnects. These contracts usually call for all electricity purchased at the full retail rate and any excess electricity delivered to the grid will be credited at an avoided cost. The avoided cost is usually similar to the cost of fuel, well below the wholesale rate and about 20 to 30% of the retail rate. This is the least desirable type of contract, but the customer usually has few choices when dealing with the local utility.

Another type of contract that is gaining in use is called "net metering." Net metering is a contract whereby any excess electricity delivered back to the utility is credited at full retail price. Most of these contracts contain a clause stating that the net electricity will be summed and evened out over some time period such as a month, a quarter, or a year. A yearly true-up provides the best savings to the wind turbine owner because excess electricity generated in one season can be used to offset deficit generation in another. As an example, high electrical use in the summer when wind speeds are low can be offset by excess generation in winter when loads are low and wind speeds are high. Monthly true-up is less desirable because wind speeds and loads do not always match up each month. Net metering rules change constantly and vary from state to state. See the web site http://dsireusa.org for the latest information on incentives for renewable energy [7]. This web site has a state-by-state listing of incentives for small wind.

Small wind machines larger than 25 kW may be asked to enter into a power purchase agreement. Again, the owner will pay full retail for any electricity consumed, but will be paid a fixed price for any electricity delivered back to the utility system. It is assumed that these larger systems may provide significant amounts of excess electricity to the grid; therefore, the utility is limiting their liability to pay larger sums for this electricity. Also, many states have capped net-metering rules to 10, 15, or 25 kW machines. This type of contract allows a better situation than the avoided cost contact for excess electricity for the bigger wind machines.

The owner needs to start early in the planning stages, meeting with the local utility and getting information on all the possible contracts for interconnection. The type of contract (permit to connect to the utility) obtained influences the profitability of the wind turbine installation greatly. This contract needs to be long term, at least 10 years, so that the owner can estimate the economic impact of his or her contract. It also needs to be renegotiable if the utility regulatory commission (state or federal) changes the contracting regulations; for example, a state utilities commission changes the rules on net metering and your system now qualifies for net metering. The owner needs to explore all possible contracts with the electric utility, as this affects the price paid for excess electricity and the price paid for any purchased electricity. A good contract can make the system a good investment, whereas a poor contract can make the system a poor investment.

REFERENCES

[1] Green J, Sagrillo M. Zoning for distributed wind power: Breaking down barriers. NREL/CP-500–38167 August 2005.
[2] Bergey M. personal communication. 2010.
[3] Preus R. personal communication. 2010.
[4] IEC 61400-2. International Standard Design Requirements for Small Wind Turbines, International Electrotechical Commission. 2nd ed. 2006.
[5] AWEA Small Wind Turbine Performance and Safety Standard, AWEA 9.1–2009.
[6] Stimmel R. In the Public Interest: How and Why to Permit for Small Wind Systems. American Wind Energy Association; 2008.
[7] DSIRE, Database of State Incentives for Renewables and Efficiency, http://dsireusa.org.

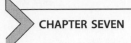

CHAPTER SEVEN

Installation
Integrating the Machine, Load, and Site into a Complete Package

Early in the discussions about purchasing a wind turbine, the installation plan needs to be considered, especially the part about who is going to do the installation. Paul Gipe [1] and other early wind enthusiasts have suggested that wind turbines with a rotor diameter less than 3 meters (10 feet) could be installed by owners. Many homeowners and hobbyists have basic construction skills that will allow them to do their own installation with these small simple systems, assuming adequate instruction manuals are provided. However, several skills are needed to complete an approved installation and it's are best done by a team of specialists. As the small wind industry has gained experience and matured, more and more trained installers are available. Also, the small wind industry, working with the Midwest Renewable Energy Association and the National Renewable Energy Laboratory, has developed a certified training program for small wind installers. Now the North American Board of Certified Energy Practitioners (NABCEP)[2] has a certification program for small wind installers. Professionals who choose to become certified by NABCEP demonstrate their competence in installing small wind systems and their commitment to uphold high standards of ethical and professional practice. Using a certified installer provides some assurance to the owner that the installer has completed training in all areas required for a successful installation.

SITE PREPARATION

The owner can do several things to get the site ready for the installation crew. One simple thing that he or she could do is to mow the site and remove tall weeds so that workers can move around the site easily. There may be several small shrubs or trees that will need to be removed. Probably the next most common items to be removed are fences that need to be moved for access to the site or, in some cases, fences that are added to keep livestock away from the wind turbine. Removal of old buildings is sometimes needed, along with clearing a path for the location of the electrical lines. It may be necessary to clear an area for the assembly of the tower and layout of materials. Clearing the work site before any actual construction begins often allows construction to proceed with few interruptions. Having a good work site makes the work go faster, without the possibility of injury.

FOUNDATIONS

The location of the tower is critical when making siting surveys and preliminary assessments of the energy potential at the site. The exact location of the base may have to be adjusted depending on the availability of an assembly location. The soil at the foundation location needs to be sampled to determine its construction properties. Loose soils such as sand often require a different type of foundation than heavier clay soils. A soils analysis should have been done during the site selection process, but is often overlooked until it is time to begin construction. Loam and clay soils will often hold their shape when drilled for several hours. Foundations in these soils can be completed with little or no forming. Figure 7.1 shows a steel reinforcing

Figure 7.1 Lowering a steel reinforcing cage into a drilled hole for a foundation for a small wind turbine (Neal).

Figure 7.2 Bolt template being placed in a reinforcing cage prior to pouring concrete for a freestanding tower foundation (Neal).

cage being lowered into a drilled hole in preparation for pouring a base for a freestanding tower. Tower orientation is often not considered to be important, but the author likes to always place the tower where the technician climbing the tower will have his back to the prevailing wind direction. The wind will push the climber into the tower instead of away from the tower. This is a small safety consideration that can help the person working on the tower feel a little more secure. Once the tower orientation is determined, then the location of the anchor bolts is determined and a template is made to hold the bolts in place while the concrete is poured (Figure 7.2). Figure 7.3 is a bolt template for one leg of a three-legged freestanding tower using a drilled pier technique. If a tower is to be erected in sandy soil, it is usually necessary to excavate a large area and form a large box-type foundation where the mass of concrete is adequate to keep the tower from overtipping. Guyed towers often work better in sandy soils because much of the overtipping load can be transferred to the guy anchors.

ELECTRICAL WIRING

Another activity that usually takes place during foundation construction is installation of the main electrical wiring. It is hoped that care was taken in

Figure 7.3 Anchor bolt template and steel reinforcing in place before pouring concrete for a single leg of a three-legged freestanding lattice tower. Pipe is for electrical wiring.

selecting the site for the tower to make sure that there was a clear path to the point where the electrical interconnection to the utility or load will be made. Typically all wiring near the wind turbine is placed in a ditch and follows the codes for buried electrical cable (Figure 7.4). All electrical wiring for small wind turbines is now covered by National Electrical Code, section 694 [3]. The new code, which was published at the end of 2011, included small wind systems for the first time. It covers all aspects of the wiring for the small wind machine from size and location of the disconnect between the wind turbine and utility, surge protection, types of raceways, and marking of all components. Even experienced installers need to review these new electrical requirements that became acceptable code at the beginning of 2012.

If the small wind system includes an inverter, that inverter must meet the UL safety standard UL 1741 [4]. Also, the interconnection to the utility must meet IEEE Standard 1547 [5], which sets guidelines for power line performance. All electrical wiring in the control panel and any disconnects should be labeled and marked clearly. Figure 7.5 shows a control panel for a small wind turbine. Many areas may require that all electrical wiring be completed by a local licensed electrician. Wind systems installed in locations outside of the United States will have similar electrical codes and regulations. Always check with local officials to make sure all electrical codes are followed and met.

Figure 7.4 Ditching for placement of electrical wiring near the wind turbine. Buried electrical cable must meet electrical codes.

ASSEMBLY OF TOWER

Typically all towers are shipped unassembled and require that space and time be allocated for assembly of the tower. Components may be shipped in several cartons or bundles, but for small wind turbines the complete tower can normally be shipped as a single truck load. If it is a monopole tower for a 50 kW or larger turbine, there may be a need for more than one truck. Some manufacturers have developed a nesting plan whereby the tubular sections are placed inside of each other and shipped as a single tube. Even with this type of shipment, there has to be sufficient room to assemble the full length of the tower. Additional space has to be provided for any motorized equipment needed to complete the assembly. Small cranes or loaders can be used to lift components and hold them in

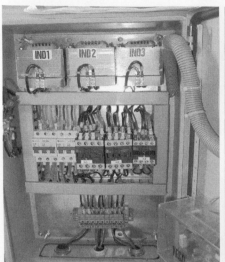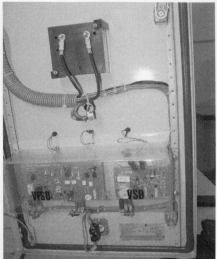

Figure 7.5 Control panel for a small wind turbine. Normally, only external connections to the wind turbine and loads are made in the control panel. All wiring must meet electrical codes.

place until fastened to the rest of the tower. All towers should be assembled on wood blocks or other supporting blocks so that adjustments can be made easily to align components. Blocks also allow access to the underneath side of the flanges for bolting and securing the assembled components.

Almost all towers will use mated flanges to connect the multiple sections. These flanges may be indexed for matching the holes so that the tower will be assembled as designed, thus meeting the design specifications. Early in the author's wind energy career, an experienced steel erector and tower assembler told him to always insert the bolts from underneath and place the nuts on top. His reasoning was that it is hard to see a nut lying on the ground if it comes loose over time, but a bolt will be found easily. Loose bolts on a tower are not found easily, but missing ones are spotted easily. Some tubular tower manufacturers use a slip joint system to connect two tower sections. This is probably more common with tubular guyed towers because the guys pull down on the upper tower section continually, always holding it in place. The slip joint may or may not be secured with bolts. Tubular or monopole towers usually require less time to assemble because there are usually only two or three flanges to connect together. Figure 7.6 shows a monopole tower almost completely assembled. The work crew is attaching the lifting arm (gin pole) to the tower. This guyed tower will be a tilt-up type of tower.

Figure 7.6 Workers are completing the assembly of a monopole tower with a gin pole. This will be a tilt-up tower.

Lattice towers require much more time to assemble because the towers are shipped with each leg separate, as well as all cross–arm and diagonal braces. Figure 7.7 shows a four–legged tower being assembled. Three of the legs have been assembled, and the braces are being assembled for attachment of the fourth leg. There are many bolts to be inserted and tightened when assembling a lattice tower. However, some small lattice towers come in preassembled 3–meter (10-feet) sections. These are assembled easily similar to the monopole towers. Before erecting any tower, make sure that all bolts have been tightened and torqued to the tower specifications. Using an

Figure 7.7 Workers assembling a four-legged lattice tower.

experienced or certified installer will make this phase of the installation go quickly and smoothly.

ASSEMBLY OF TURBINE

Generally the nacelle is shipped completely assembled, but with some larger machines, a hub may have to be attached at the erection site. Assuming that the nacelle has the interior components assembled and that the internal wiring is completed, the nacelle is ready to be mounted on the tower. Again, for most machines, when the assembled nacelle and rotor weigh less than 500 kg (1,100 lb), they can be attached to the tower on the ground and the complete unit can be lifted altogether. If a tilt-up tower is used, the rotor blades are attached to the hub or PMA magnet can at ground level as well. Figure 7.8 shows workers attaching rotor blades to a PMA magnet can while the unit is on the ground. Most installers prefer to assemble as much of the system on the ground as possible rather than working from the tower or a man-lift. Care should always be exercised when attaching rotor blades to make sure they are attached for proper rotational direction. Not all manufacturers use the same rotational direction, and blades have been attached for improper rotation direction. Also, someone may attempt to install one blade opposite the others. Again, an experienced installer does not make these mistakes. Care should be taken to make sure all blades are set at the same pitch. Pitch is the twist from the

Figure 7.8 Attaching rotor blades to a wind turbine generator before the complete unit is lifted into place.

normal plane of rotation with respect to the center line of the blades. An increase in the pitch angle may increase performance and output more power than the generator is capable of handling, whereas a decrease in the pitch angle may make the machine underperform and not reach the expected output. Several manufacturers use shims under the blade attachments to adjust the pitch. Typically, the proper pitch angle is determined by computer simulation for each site using available wind speed data. This simply means that not all machines of the same model will have the same pitch setting when installed because of adjustments to maximize energy production at each site.

After the wind turbine nacelles are fully assembled with rotor blades and all tower top wiring completed, they are ready for lifting into place. For freestanding towers and guyed towers without a tilt–up system, a crane is used to lift the entire unit into place. The only difference between these systems and ones on a tilt–up tower is that a crane is not necessary to lift tilt–up towers into place. However, as the turbine size increases and the nacelle weight is above 500 kg (1,100 lb), the tower is typically erected and secured in place before installing the nacelle. This procedure is more like installing a utility size wind turbine in that the nacelle, rotor blades, and tower are erected in separate steps. Figure 7.9 shows a second rotor blade being

Figure 7.9 Attaching a second blade to a nacelle in preparation for installation on a tower. Larger turbines often require erection of subcomponents because of size and weight.

Figure 7.10 A nacelle with two blades attached is ready to be lifted for attachment to the tower. The third blade will be added in another step.

attached to a nacelle while still on the ground. Figure 7.10 shows the nacelle being lifted with two blades attached. After attaching the nacelle to the tower top, the third blade is attached. Some installers prefer to attach the third blade with the blade pointing down by hoisting the blade up into position, whereas other installers prefer to attach the third blade with it pointing up and allowing the weight of the blade to hold it against the mounting flanges. The author has participated in installing blades both ways and has found that it is mostly dependent on the skill of the crane operator as to which method is best. Again, with these larger machines, it is best to use an experienced installation crew.

A part of turbine assembly is the electrical wiring at the tower top. All electrical wiring should be completed by an experienced electrician who knows how to secure wires and protect them from damage by wind, rain, ice, etc. Wiring at the tower top is exposed to all types of environmental hazards and should be protected as much as possible. Another hazard to exposed wires is the removal of insulation caused by birds picking at the wire, especially small control wires. All connections should be made inside of an approved weather-tight enclosure and control wires placed in a flexible conduit. Wiring going down the tower needs to be supported so that all the wire weight is not on the electrical connection but on the cable itself. Wires going down the tower to the ground need to be fastened securely every 2 to 3 meters (6 to 10 feet) to prevent flopping against the tower, wearing the wire insulation, and causing an electrical short. An exception to this is if the manufacturer provides a single heavy cable with multiple conductors that allows the cable to twist as the turbine yaws. This type of cable is used in place of using slip rings, which allows for the turbine to yaw as many times as necessary. Twist cables require maintenance semiannually, but because they cost much less than slip rings they are popular in small wind turbines.

LIFTING THE ASSEMBLED COMPONENTS

After all components have been assembled, the unit should be ready to be erected. The procedures to be used vary depending on the type of tower used and the size of the unit. Again, for units that are to be on a tilt–up tower, the procedure is simple. It is assumed that the lifting of the tilt–up tower was practiced and checked before attaching the wind turbine. It is always best to check the operation of the tilting mechanisms before attaching the turbine. The completed unit is assembled, and the turbine rotor is secured by a brake that can be released at ground level or is furled so that it will not begin rotating during the lifting process. Once the tower is vertical, it needs to be secured by turnbuckles and all pulleys relieved of tension. The next step is to adjust the tension in all guy wires and make sure the turbine is leveled. It is usually assumed that the top mounting plate is perpendicular to the tower so that if the tower is exactly vertical, the mounting plate is level. If a turbine does not appear to yaw properly, the levelness of the tower top will need to be checked, which often requires someone to access the tower top to measure the levelness. This is not a common problem, but has happened with newer, prototype machines. The turbine should be set ready for operation when the tower is leveled and all guy cables are tensioned to the specified tension.

The erection of freestanding and nontilt–up guyed towers is similar if the turbine is of the smaller size. Assuming that space is available to assemble the entire length of the tower, the unit can be lifted in a single lift. The assembled wind turbine is attached to the tower and all wiring is completed. The rotor blades need to be secured to prevent rotation during the lifting process. Units without brakes need to have one of the blades tied to the tower to make sure that no rotation of the rotor occurs, as the blade might strike the lifting cable and cause damage to the rotor blade. A lifting crane is attached to the wind turbine or tower top and is positioned so that it lifts the

Figure 7.11 Lifting a small wind turbine and tower as one lift using a crane. A tractor with a front-end loader was used to hold the base of the tower off the ground while lifting the tower.

unit straight up. Another portable lifting device is positioned at the lower tower section or base section. It is used to lift the base of the tower and keep it from dragging on the ground as the tower is lifted into a vertical position (Figure 7.11). Once the tower is vertical, the portable lifting device is removed and then the tower can be positioned over the foundation. All bolts need to be tightened securely and all guy wires tensioned properly before the turbine is released for operation. Someone will need to access the tower top to release the crane from the tower. No one should attempt to climb the tower before it is fastened securely and ready for operation. With these types of towers, final wiring at the tower base has to be completed before operating the turbine.

Turbines larger than 25 kW often require that the tower be mounted in place before attaching the turbine. As discussed earlier, many times the third blade may have to be added after erecting the nacelle. Installation and erection of these machines follow more of the process of installing large utility-type machines. The process has several steps instead of just one or two like the smaller machines.

INITIAL OPERATION OF THE UNIT

After all erection is completed and wiring is completed, it is time to prepare for the initial operation. The owner's manual should contain information about what checks should be made as you begin this phase. Follow the instructions completely. It is best to release the brake and let the rotor turn slowly and reset the brake. It is important to make sure that the brakes are capable of stopping the turbine before you let it run to full operating speed. If the machine has an induction generator wired for three-phase operation, it is important to check for proper rotation direction. Three-phase systems can be wired to operate in either direction, and it is impossible to determine the direction without trying it. It is best to do the initial operation in light to moderate wind speeds so that the turbine rotor will not accelerate to maximum speed rapidly. Once the turbine starts to produce electric power, it is best to check for proper voltage, frequency, and current. These measurements can be made with portable measuring devices if the turbine does not include instruments for measuring them continuously. If everything checks out, then the machine can be left running. Installation is completed.

COMPLETING THE SITE

Depending on the location of the wind turbine, it may be necessary to provide various levels of security. It is always a good idea to secure the control panel for the turbine by either locking the panel or placing it in a building. Some sites may require fencing to provide security and limit access to the tower base. If the tower is climbable, then it may be necessary to remove the lower portion of the climbing ladder or secure it to limit climbing. All electrical panels should be marked with warning signs about high voltage and electrical shock. To complete installation, the site needs to cleaned of all erection equipment and shipping cartons and put in an orderly fashion.

REFERENCES

[1] Gipe P. Wind Power, Renewable Energy for Home, Farm, and Business. White River Junction, VT: Chelsea Green Publishing; 2004.
[2] NABCEP, North American Board of Certified Energy Practitioners, www.nabcep.org.
[3] National Electric Code, 2011 edition, National Fire Protection Association, www.nfpa.org. NFPA 70.
[4] UL Standard 1741, Inverters, Converters, Controllers and Interconnection System Equipment for Use with Distributed Energy Resources, http://ulstandardsinfonet.ul.com.
[5] IEEE Standard 1547, Interconnecting Distributed Resources with Electric Power Systems, www.ieee.org.

CHAPTER EIGHT

System Operation with Electrical Interconnections

Working with Electrical Utilities and Transmission Companies

Connecting a small wind turbine to an electric utility is composed of two different but necessary components. One is the policy and permitting process and the other is the actual physical connection. One is doing and processing all the paperwork to get permission to connect and the other is overcoming the technical issues and making sure the connection is dependable and safe. A general description of the permitting process is discussed in the chapter on permitting, but this information will help you understand the issues when dealing with the different types of electric utilities.

Understanding the electrical grid in any country requires study and asking lots of questions for most small wind turbine purchasers. Many countries have a government-owned utility that serves the entire country, which means that there is a uniform policy throughout the country. However, in the United States, there are many utilities and each state has the authority to establish policies for their state. The U.S. Federal Energy Regulatory Commission is charged with approving rates for wholesale sales of electricity and transmission in interstate commerce for jurisdictional utilities, power marketers, power pools, power exchanges, and independent system operators [1]. They also do certification of qualifying small power production and cogeneration facilities. Each state has a utility regulatory commission that establishes policies and rates and controls some transmission of electricity. Most of the regulatory power resides with the states because they manage the rates of retail electricity. Another issue with the U.S. electric market is that it is divided into three different electrical grids (Figure 8.1). The three grids operate totally independently of each other because the alternating cycles are not in phase. The western grid is 180° out of phase with the eastern grid, and the Electric Reliability Council of Texas grid is 90° out of phase with both eastern and western grids. The only

Small Wind
ISBN 978-0-12-385999-0

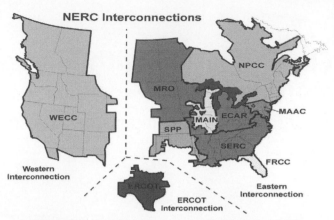

Figure 8.1 Electric power grids of North America [1].

transfer points between grids require a switching system whereby alternating (AC) electricity from one grid is converted to direct current and then converted back to AC to match the receiving grid. Typically, these switches are small and do not allow for large amounts of power to be transmitted at one time. Note that because Canada is also included with the United States in these national grids, it is easier to move electric power north and south than east or west.

The United States has electric utilities that are generators of electric power. They own and operate a number of generating plants and sell their power on the wholesale market. Most large wind farms or wind plants would be considered a generating company only. They depend on companies that are mainly transmission companies to move their power to lucrative markets. Transmission companies exist to move electric power from one location to another to meet electrical demands and satisfy consumer's hunger for more electricity. Finally, some electric companies are distributors. They provide service to individual users and collect for actual use of electricity. These companies collect the retail rates for electricity and provide income for the generators and transmission companies. Some large electric utility companies actually do all three components of the electric business by being a generator, providing transmission lines, and doing retail distribution.

Almost all small wind turbine systems are connected to a local retail distribution system; therefore, the local utility establishes the interconnection policies and equipment requirements. It is important to understand the type of local retail electric company you are working with. Local utilities are usually investor owned, municipal owned, or a cooperative. Investor-owned companies are owned by investors or a corporation of investors who are providing a service with the intent of making a profit for the investors. These companies are usually large multistate companies or have a large number of retail customers. They typically are a combination of generation, transmission, and retail sales. Because their goal is to make a profit, they typically are leaders in innovation and development of new technologies for the production and marketing of electricity. Many of these companies have already established policies and procedures for inter-connecting small wind turbines into their electric grids. Fees and rate structures for independent generators have been established and published as part of their business polices approved by the state energy commissions.

Municipal-owned utilities are owned and operated by individual cities. These systems were established many years ago when electricity was first developed for cities. Many of these systems have been sold to an investor-owned utility since the early 1990s because cities no longer found it feasible to own and operate a small utility. Also, the capital cost for a new generation capacity was an issue for many cities facing rapid growth. Today, many municipal utilities purchase much of the electricity sold within the cities. As a result, these municipal utilities are retailers only of electricity and often have clauses in their wholesale purchase agreements about how much locally generated electricity can be added to their systems. They may even have to submit a request to their wholesale provider for each third-party generator they allow to connect to their system. Therefore, connection to a municipal system may require several approval steps.

Another type of local electric utility that small wind systems will be connected to is rural electric cooperatives. These systems are owned by the individuals who purchase electricity from the system. They were established under the guidelines and orders of the U.S.D.A. Rural Electric Adminis-tration. These electric cooperatives would be classified as retail-only utilities because very few have generating capabilities. They usually purchase power from either generating companies or, in some cases, from transmission companies. As with municipal utilities, they may or may not have clauses in their purchase contracts about allowing third-party generation on their distribution lines. Because changes in policies have to be voted on by the cooperative membership, new policies to allow for wind turbine inter-connections have been approved slowly.

Most cooperatives in windy areas have had to address the issue because of member request, but others in marginal wind areas have not considered interconnection policy changes. Again, approval for connection to a cooperative utility may require several steps and considerable time and effort.

Begin early in the planning process by contacting your local electric provider. Also, check for information at the state public electric commission. State commissions will have information about buy back policies and any rules about net metering. Net metering is a concept whereby the utility will allow the wind turbine operator to feed back a portion of the electricity at the same price as is paid for the electricity. Net-metering rules and policies vary greatly from state to state and not all states allow net metering. The web site http://Dsireusa.org has information about policies and regulations related to renewable energy for all 50 states. However, for detailed information about how your local utility is going to address your request for interconnecting, you need to work with the person responsible for receiving and handling your request for interconnection. Be well informed concerning the policies and requirements before making a special request from the utility.

CONNECTING THE WIND TURBINE

Once the permit to connect to the utility is granted, it is time to begin considering the physical connection. All small machines use a simplified interconnection; these connections can be as simple as plugging the line into a wall outlet if the voltage and amperage are below 230 volts and 20 amps. However, the author does not recommend this practice because of potential electrical shorting from voltage spikes caused by lightning and equipment operating on the electrical system. It is recommended that a fused disconnect with lightning surge protection be used between the wind turbine and the electric grid. This allows for the turbine to be disconnected from the electric grid during maintenance and repairs to the wind turbine. The fused disconnect protects both the wind turbine and the grid from overvoltage spikes that sometimes occur on electrical systems. This fused disconnect is not to be confused with a disconnect that is usually required at the metering point.

Almost all utilities require that the metering point be equipped with provisions for measuring the electric flow in both directions, that is, the incoming electricity that is metered for retail pricing and the outgoing electricity that is metered for wholesale or avoided cost. Traditionally, two meters have been used at these installations (Figure 8.2), but in recent years,

Figure 8.2 Typical electric meter installation for a small wind turbine. The top meter measures electricity used from the utility, and the lower meter measures excess electricity fed back to the utility. Note the line disconnects in the background.

the new smart meters are capable of measuring the flow of electricity in both directions. The utility will specify the type of metering desired by them for your application. Utilities also require a disconnect at this point so that their service personnel can disconnect the service with an independent generator to ensure that the line is free of electric energy. This protects personnel during repairs and maintenance on the utility lines near the wind turbine generators. The author adds that the owner is expected to pay all the costs for any extra equipment that is required or desired by the utility. This will normally include special wiring for the meter, the meter housings, and the disconnect. Do not overlook these electrical costs when estimating the installation costs.

Again, all electrical wiring should be completed in a manner that meets the electrical wiring codes for your location. As a minimum, it should meet Section 694 of the U.S. National Electrical Code. Using a certified small wind installer ensures that all current electrical codes are followed and that you will have an approved electrical connection to the utility and your loads. Almost all nations now have some national electrical wiring code that should be followed. Cities and counties often have additional restrictions that must be followed, as well as national codes. All of these codes are used to protect owners, repairmen, and the equipment that is being installed. Be sure to use qualified workers to do the installations and have an independent inspector check the wiring before energizing the system.

OPERATING THE WIND TURBINE CONNECTED TO THE GRID

Many electric utilities offer net metering for small wind turbine operators. In the United States, 43 states offer some form of net metering [2]. These range from actually giving the wind turbine owner full retail credit for excess electricity generated to paying an avoided cost for electricity fed back into the grid. Table 8.1 shows the monthly purchase rates for renewable energy systems in New Mexico for the fall of 2011 [3]. They offer an on–peak rate and an off–peak rate where on peak is usually about four hours in length during the late afternoon and early evening; typically 4 to 8 p.m. These purchase rates are about 25% of the retail rate, which varies depending on the cost of fuel for the coal and gas electric generators.

Many states limit the size of generator to 10 or 25 kW for net metering; any larger wind turbines require a power purchase agreement type of contract. In almost all cases, it is best to manage your electrical use where you consume most of the electricity you generated and sell as little to the utility as possible. Remember, the main purpose for having a small wind turbine is to offset or avoid higher electric utility bills. You cannot make a small system economical by selling a large portion of the generated power to the utility at low avoided costs. You must use the power in place of purchasing power from the utility.

Almost all net–metering programs use one of two "true–up" systems. The majority of net–metering programs balance the amount consumed

Table 8.1 Monthly energy purchase payments to qualifying facilities in New Mexico, effective August 21, 2011

Month	On-peak rate per kWh ($)	Off-peak rate per kWh ($)
January	0.0499	0.0372
February	0.0393	0.0371
March	0.0269	0.0288
April	0.0402	0.0284
May	0.0402	0.0434
June	0.0321	0.0255
July	0.0385	0.0349
August	0.0344	0.0319
September	0.0304	0.0245
October	0.0262	0.0207
November	0.0362	0.0319
December	0.0341	0.0283

against the excess produced each month. This is a monthly "true-up" program [4]. This often means that there are a few months that have good winds and low heating or cooling demands, which means less electric usage. These months will sometimes have more electricity generated than consumed. However, there are more months when heating and cooling demands are high and wind may not be as good as spring and fall months, thus requiring that you still pay a large electric bill. Monthly true-up does not allow you as a customer to produce excess power in one season and use it in another. However, a few net-metering programs allow for yearly true-up. This program is much more customer-friendly in that it allows for excess electricity to be credited against periods of electrical shortage. The type of net-metering program you have will influence how you manage your wind system and the times that you will perform many tasks.

Business owners need to evaluate their workloads carefully to determine if they can be modified to match the daily and seasonal wind speed changes. Some areas have daily wind-speed cycles that can be matched with daily work activities. Other locations have a large seasonal variation with high fall and winter winds and low spring and summer winds. As a business owner, can you reduce your output in periods of low winds and increase your output in seasons of high winds? These management decisions will influence the success of your wind project greatly.

REFERENCES

[1] Federal Energy Regulatory Commission (FERC) web site: www.ferc.gov/industries/electric.
[2] DSIRE Web site: http://dsireusa.org/incentives.
[3] Public Service Company of New Mexico (PNM) web site: www.pnm.com/customers/interconnection.
[4] National Grid Web site: www.nationalgridus.com/niagaramohawk.

CHAPTER NINE

System Operations of Stand-Alone Machines

Making Systems Work without an Electrical Grid

Most of the uses of small wind machines for years have been systems that operate in a stand-alone mode. The interface of a wind turbine to an electrical grid is a relatively new concept. It has only been since about 1970 that wind systems have been connected to an electrical grid in large numbers. Wind units for grinding grain, pumping water, and other similar uses have been used for several centuries, dating back to development by the Persians [1]. Small windmills with several sails were used all around the Mediterranean Sea during 1500s to 1900s to pump water and grind grain. The Dutch developed two types of wind machines that would be classed as small wind machines for pumping water and grinding grain in the 1600s and 1700s (Figure 9.1). These units could be yawed into the wind by using a lever or pulley system and then locked into place. These technologies were brought to the Americas when immigrants moved from Europe to America. However, much of America did not have the one-directional, smooth wind speeds that were found in Europe. In 1854, Daniel Holladay patented the design that would evolve into the American windmill [2]. The main purpose of the American windmill was to pump water from a well rather than pumping water from one canal to another canal as with the European systems. Figure 9.2 shows a picture of an Eclipse wooden wheel windmill built from 1873 to about 1920. In the 1920s, manufacturers begin making the wheels out of metal and developed an enclosed self-oiling gearbox. The Aermotor Windmill Company has been making windmills since 1888 and continues to manufacture about 1,000 units per year and about another 1,000 replacement gearboxes and wheels [3].

The first electrical units manufactured for charging batteries used small DC generators that produced 6 to 32 volts. Jacobs and Windcharger were the two significant manufacturers; however, mail order companies such as Sears Roebuck and Montgomery Ward sold thousands of units in the 1930s and 1940s. Both Jacobs and Windcharger stopped production in 1956 [4].

Small Wind
ISBN 978-0-12-385999-0

Figure 9.1 This Dutch-style water-pumping windmill built in the 1600–1700s was used to remove flood waters from cropland.

These units provided enough power to operate one or two lights for three to four hours and power for a radio. Figure 9.3 shows a 200 watt Windcharger used for charging a 12 volt battery.

Water pumping, grain grinding, and small DC electrical systems are the historical uses of small stand-alone systems. Today, stand-alone systems are still the largest uses for the smaller wind turbines purchased. Worldwide, almost 75% of the systems sold are for off-grid or stand-alone applications and are all smaller than 10 kW [5].

BATTERY CHARGING

The world market produces many small wind systems less than 1 kW in size and almost all of them are used for battery charging. Figure 9.4 is

Figure 9.2 The Eclipse windmill, first built in 1873, was the beginning of a design that became known as the American windmill.

a schematic of a small stand–alone battery-charging system. It is composed of a wind turbine with a DC generator, batteries, and a charge controller. The system is connected directly to the on–off switches for the loads. The charge controller is a key component in this system because it limits the current flow to the batteries. As long as the batteries need additional charging, then the controller allows current to flow to the batteries, but once the batteries are fully charged, it shuts off the current flow to the batteries. Some systems may supply an auxiliary or dump load with power when batteries are charged, whereas others will simply open the circuit between the wind turbine generator and the batteries. When the circuit is opened, the wind turbine must have an overspeed control that will allow the turbine to run unloaded in all winds. Many of these small units will furl out of the wind and cause the rotor to slow rather than apply some type of brake. The reason that it is important to have a charge controller in the system is to prevent

Figure 9.3 A 200 watt Windcharger, an early electrical-generating machine used to charge batteries, was built from the 1930s until 1956.

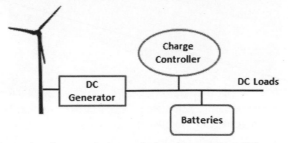

Figure 9.4 Schematic of a stand-alone wind turbine with a DC generator used for battery charging.

overcharging the battery or batteries. Overcharging heats the batteries and boils the liquid out of the battery and shortens the lifetime of the battery.

In recent years, fewer battery–charging wind turbine manufacturers are using DC generators and more are using a permanent magnet alternator (PMA) with a rectifier to obtain the DC output needed for battery charging (Figure 9.5). The advantage of the PMA is that it allows operation over a wider range of rotor speeds and efficiency is not lost with either high rotational speeds or low rotational speeds. The alternators also have a longer lifetime and less maintenance than DC generators.

The many examples of battery–charging systems range from using them in remote areas where no utility grid is available for home power to using them in water applications for safety-warning devices and telecommunications.

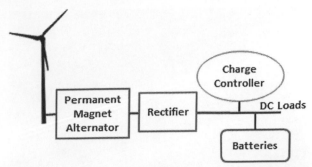

Figure 9.5 Schematic of a stand-alone wind turbine with a permanent magnet alternator used for battery charging.

One of the largest uses is for systems that are erected easily on tilt–up type towers and are moved easily from location to location. China is the largest manufacturer of these units and they are for the nomadic peoples of west and northwest China who still reside in tents. These systems provide power for a few lights and either a radio or a television. In recent years, the size of these units has increased and is now averaging about 400 watts. China exports some of these small systems all across Asia and the Middle East.

Another major use of these small systems is for keeping batteries charged on sailing ships (yachts) and houseboats (Figure 9.6). These boats are equipped with 12 volt DC appliances, lights, and other electrical accessories. The engine is often not operated enough to keep batteries charged and using a small wind turbine is a good alternative for keeping batteries ready for use. Typical turbine size is 200 to 400 watts with a DC output. The towers or mounting mast for these units is usually just tall enough to clear everyone's head and hands. There is usually sufficient wind to keep batteries charged with the limited use of the onboard electrical items.

In recent years, small wind systems have been installed to provide security and safety lighting along highways and in large parking lots. Figure 9.7 shows a small wind turbine powering a street light and a flashing red stoplight at a highway intersection. These systems are often very cost-effective because of the cost involved of extending a grid line to the location. It costs approximately $10 per meter for a low power grid extension that would operate a security light and blinking light.

Still another application of small battery–charging systems is being used in less developed countries where there are many villages without any electrical grid system. Wind machines are being used to develop battery–charging stations where several batteries are recharged at the same time. Homeowners

Figure 9.6 Wind turbine used to charge batteries on a yacht (NREL).

bring their batteries to the charging station and either leave them for recharging or exchange their old discharged battery for one that is fully charged. They take the fully charged battery home, attach it to their home circuits, and use it until it is discharged. They pay a small fee each time they exchange a battery, which provides an income for the battery-charging operator. Most of these systems are government-sponsored programs to provide a small amount of electrical power in remote villages.

There are still a large number of remote homes that are off the electrical grid that use battery-powered lights and appliances. Many of these homes have installed small wind systems to keep the batteries charged. Other homeowners have selected to just disconnect from the grid and do their own generation. The availability of modern, efficient 12 volt appliances developed for motor home, travel trailer, and boating industries has provided

Figure 9.7 Wind turbine used to charge batteries for security lighting and traffic control at a remote highway intersection.

remote homeowners with alternatives to grid electric power. Many of these homes have also installed solar panels to provide additional power and a dual-energy source. These hybrid systems are described later in the chapter.

WATER PUMPING

Historically, grinding grain and water pumping have been the two most important applications of wind power. After the introduction and widespread use of the steam engine and later the internal combustion engine, grain grinding switched rapidly from wind to these more controllable power sources. However, wind power remained a principal power source for lots of water pumping because many of the pumping sites were remote. Europe used wind-powered pumps for flood control and moving water from one surface body to another. The Cretes used wind power for pumping from shallow wells, replacing animal power. Shallow wells are usually less than 10 meters deep. The American windmill was the first wind-powered system to be designed specifically for deep well pumping. Initially they were designed for wells less than 60 meters deep, but it is not uncommon for wells to approach 100 meters today. Figure 9.8 shows an American windmill providing water for range cattle. This is the most common use of the windmill today, but the windmill allowed settlers to live

Figure 9.8 American windmill used for pumping water for livestock (same design used since the 1930s).

away from a continuously running stream beginning in the 1880s. Most of the Great Plains region of the United States was not settled until after the development of the windmill. This is also true for much of Australia, Argentina, and South Africa. For almost 40 years from 1880 to 1920, the City of Amarillo, Texas had windmills in town at each house.

WIND–MECHANICAL PUMPING

The American windmill was designed to pump water from a drilled well with a bore hole of 15 cm (6 inches) to 20 cm (8 inches) in diameter. The windmill tower was centered over the well and the well pipe was inserted below the water line (Figure 9.9). A movable rod was attached to the windmill head at the top and a cylinder pump or piston pump was attached at the bottom. As the rod moved up and down, water was lifted to the surface. The cylinder pump consists of three components—a pump barrel, a nonmoving check valve, and a moving plunger (Figure 9.10). The check valve and plunger have cups made of leather or neoprene that seal against the pump barrel. The pumping capacities of four common cylinder sizes are listed in Table 9.1 along with the lift elevations for different sizes of wind rotors. The two smaller cylinder sizes are the most common for livestock water because they provide sufficient capacity and work with common drop pipe sizes. The 4.8 cm pump is a $1\left\{\dfrac{7}{8}\right\}$ inch diameter that

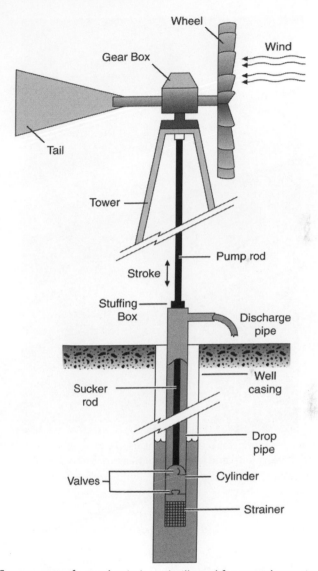

Figure 9.9 Components of a mechanical windmill used for pumping water from wells.

screws onto a 2 inch pipe. This combination is used because the valves can easily pass through the drop pipe and then seat into the cylinder and the cups will seal against the cylinder walls. Each cylinder size is matched to a standard pipe size, which can be used as the drop pipe.

The American farm windmill is characterized by a high-solidity rotor consisting of 15 to 18 slightly curved blades or vanes. The large number of

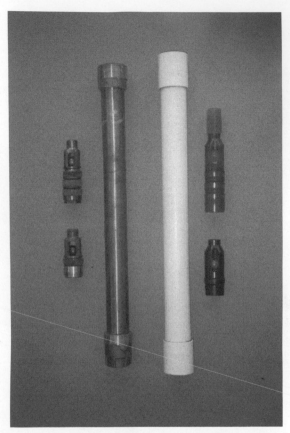

Figure 9.10 Cylinder pumps or piston pumps used with mechanical windmills. Pumps are constructed of brass (left) and from PVC with a urethane/epoxy lining and urethane check valves (right).

Table 9.1 Pumping capacities and lift elevations for multiblade windmills using piston pumps

Cylinder size (cm)	Pumping capacity (m³/hr)	Rotor diameter (m)				
		2.44	3.05	3.66	4.27	4.88
		Elevation (m)				
4.8	0.68	53	79	119	171	280
5.7	0.98	34	52	76	110	180
7.0	1.45	24	37	55	79	129
9.5	2.76	—	20	30	44	70

blades that make up the rotor or wheel provides a high starting torque that is needed for operating the piston pump. Most units have a back gearing system that transfers the rotating motion of the rotor to a reciprocating motion for pumping water. The wheel and tail are usually made of galvanized steel and all components are bolted together. This makes replacement of damaged parts relatively easy to do in the field. Towers are normally made of steel, and most companies offer a tower top that allows for mounting on a wooden tower. Pump rods can be made of wood, steel, or fiberglass. Almost all of the pump cylinders are made of brass, but newer PVC pumps with a urethane/epoxy lining and urethane check valves are being used.

The volume of water pumped is controlled by adjusting the number of strokes per minute, which is controlled by spring tension on the tail. The tension is adjusted so that maximum strokes per minute range between 30 and 35 strokes per minute. Most machines reach the maximum stroke speed at a wind speed of 9 to 10 m/s and thus reach the maximum pumping rate at that same wind speed. In higher winds, the rotor furls out of the wind and the rotor slows so that strokes per minute slow and the pumping rate drops. The rotor for the American windmill has a peak power coefficient (C_p) of about 30% at a tip–speed ratio of around 0.8 [6]. The efficiency for the reciprocating piston-type pump is essentially constant at 80% over the operating rage of wind speeds. The overall annual efficiency of the system is between 5 and 6%.

The American farm windmill has a long life, 40 to 50 years; however, the main problem is that the leathers or cups on the pump have to be changed on a regular basis, usually once a year depending on the water quality. The cost of pulling the rods is around $200 if a well rig is used and if done with hand labor, approximately one man day. The other major maintenance is to change the oil in the gearbox once per year. This expense, along with the fact that many American farm windmills are at the end of their lives, brings an opportunity for new types of wind and/or solar pumping systems.

WIND–ELECTRIC PUMPING

The wind–electric water pumping system consists of a wind turbine generator connected directly to a standard induction motor driving a centrifugal or submersible turbine pump (Figure 9.11). There is a good match between the wind turbine output and the centrifugal pump because both have power proportional to the rpm cubed. Wind–electric pumping systems have shown a higher overall efficiency than mechanical systems, and

Figure 9.11 An electrical-generating wind turbine providing electricity directly to an electric pump for pumping water without an electrical grid.

because systems with PMA are typically larger than mechanical systems, more water can be pumped with electrical systems. Another advantage of wind–electric systems is that the wind turbine can be located some distance from the well or pump.

Wind–electric water pumping systems were developed by the U.S. Department of Agriculture (USDA)–Agricultural Research Service (ARS),

Conservation and Production Research Laboratory in Bushland, Texas in cooperation with the Alternative Energy Institute, West Texas A&M University, Canyon, Texas. They experimented with several different wind turbine sizes and different motor pump combinations before finding a working system. Output from the PMA had a variable frequency proportional to the speed of the wind turbine rotor. As the rotor speed increased, the frequency increased almost linearly. Voltage also increased as rotor speed increased, but not in the same proportion as the frequency. A wind turbine controller was designed that would add capacitors to the circuit to boost the voltage when the frequency exceeded 60 Hz (Figure 9.12). The controller managed the voltage–frequency ratio to maintain it near 3.7, the ratio for an induction motor operating at 220 volts and 60 Hz. The system would pump water over a voltage range of 170 to 270 volts and frequencies of 35 to 70 Hz [7]. A dump or dummy load was added to the system so that excess energy could be diverted to it when winds exceeded the power required for the pump motor (Figure 9.13).

Several different wind turbines ranging from 850 W to 10 kW were used in testing these pumping systems. Figure 9.14 shows a pumping curve for a 1,500 watt wind turbine with a 3.05 meter rotor diameter lifting water from a 73 meter (240 feet) deep well [8]. Also shown in Figure 9.14 is a pumping curve for a multiblade mechanical windmill with the same rotor diameter. Note that the electrical system requires a higher start–up speed, but

Figure 9.12 Controller for a wind–electric water-pumping system designed by the USDA-ARS in cooperation with the Alternative Energy Institute, West Texas A&M University at Bushland, Texas.

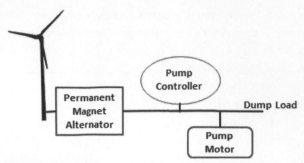

Figure 9.13 Schematic of a wind–electric water-pumping system. Excess wind power is diverted to the dump load when too much wind power is supplied for the pump.

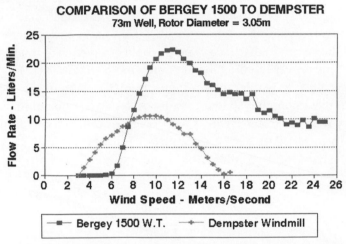

Figure 9.14 Pumping curves for a wind–mechanical (Dempster) and a wind–electrical (Bergey 1500) wind turbine pumping water from the same pumping depth. The two wind systems have the same rotor diameter.

pumps more than twice the volume of water than the mechanical windmill. Another comparison is done using similar pumping curves and a Rayleigh wind speed distribution and calculating daily water pumped in each month for both water pumping systems. This helps show water available for domestic use or to water livestock on a daily basis. It can also be used to determine if sufficient water is available for watering a small vegetable garden or a small farm plot. Figure 9.15 shows daily water pumped for a pumping lift of 45 meters (150 feet). Note that the mechanical system pumps about the same amount of water in each month because it does not fully utilize wind speeds above 10 m/s. However, the electric system

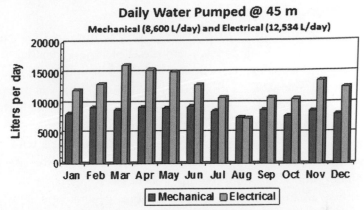

Figure 9.15 Comparison of daily water pumped in each month by a wind–mechanical windmill and a wind–electrical wind turbine. Both systems pumped from the same pumping depth.

continues pumping at the higher wind speeds and pumps more water in all months except August when wind speeds are lowest.

Small wind systems are a great energy source for pumping water in remote areas or in areas where wind is sufficient and grid electricity is expensive. Several systems are available for purchase, either mechanical or electrical. A well driller or water well repairman in your area can help you select a pump suitable for your wind speed conditions and provide the skills to help with installation and maintenance.

HYBRID SYSTEMS

Another type of stand-alone application for small wind systems is when wind-generating systems are combined with other electric-generating technologies. The most common are combining wind and solar photo-voltaic systems for battery charging and/or with use with a small inverter for producing AC-type electricity. Many times these systems are not connected to the electric grid and therefore operate as a stand-alone system. The other type of hybrid system combines a wind system with a petroleum-fueled engine powering an electric generator. The most common fuel for these engine-powered generators is diesel because of its their reliability when running for long hours at relatively low speeds. In very remote sites, some hybrid systems use all three energy sources with a diesel generator as backup in case of failure of one of the two primary sources of wind and solar.

Hybrid Systems for Telecommunications

The need for continuous electric power at very remote sites launched the development of hybrid wind, solar, and diesel electric-generating systems. The Department of Energy teamed with some private companies to develop these hybrid systems. Work with Northern Power Systems was successful in developing a small reliable system that was installed at Antarctica in the early 1980s. This project was so successful that several hundred remote sites were developed in the next 20 years. These sites contained a reliable wind turbine designed for extreme winds and often very cold temperatures along with photovoltaic panels all connected to a battery storage system. A diesel generator was included at the site as a backup generating system in case batteries needed additional charging. It is unknown where all these systems were installed, as many of them were purchased by the military and installed in secret locations for security purposes.

The need for these remote telecommunication systems has diminished in the last 5 to 10 years with the use of satellites, cell phones, and the Internet. Now, almost anyone can have access to world information through the Internet and network services via satellites. However, this technology development was very important for developing vital applications for small wind systems and it proved that wind machines could be built that are reliable and will operate in almost any environment.

Wind–Solar Hybrid Systems for Homes and Cabins

Around the world, many families own small vacation homes or cabins in mostly remote areas or islands where reliable electric grid power is not available, available for only a few hours per day, or extremely expensive. With some of the advances in battery technology and DC appliances, generating systems for charging batteries and operating a remote cabin can provide many of the same conveniences that people are accustomed to at their city residences. By combining both wind and solar systems together, the possibility of not running out of electricity is improved greatly. Observations recorded on the operation of wind systems show that a wind system in an area with an average wind resource of 5 m/s will actually produce power about 60 to 65% of the time. A solar photovoltaic array will produce power on an average of about six to seven hours per day in the winter months and 11 to 12 hours in the summer. A good average is about 35 to 40% of the time for the solar system. When these two energy sources are combined, they produce power 80 to 85% of the time, depending on nighttime winds mainly in the summer months [9].

Because many small wind installers also sell and install solar systems, this combination is easy for them to do. There are several good sources of information to help an individual determine the relative size of wind turbine and photovoltaic array to create a wind solar hybrid system. Paul Gipe gives examples of sizing wind–solar hybrids in his book *Wind Power, Renewable Energy for Home, Farm and Business* for both a cabin-sized system and a household-sized system [10]. In both cases the solar array is about half the size of the wind system and the battery storage is almost six times the generating capacity. Richard Perez, publisher of *Home Power* magazine, also has guidelines available for sizing wind–solar hybrid systems for homes [11]. Oftentimes, homeowners will also purchase a diesel generator for backup in case one of the systems fails due to extreme weather conditions or other unexpected occurrences. They may or may not be connected to the system for automatic start-up depending on how important it is to always have power available. Also, the standby generator may be smaller than the wind or solar system because it may be used to ensure operation of some essential items such as a refrigerator or to provide heat to prevent pipes from freezing in winter.

These small cabin or household hybrid systems can be purchased and installed in modular fashion by first purchasing the wind turbine, hybrid controller, inverter, and some batteries. Solar panels and even additional batteries can be added later. Solar panels and batteries can be purchased in small increments and added as needed or desired. However, the wind turbine has to be purchased as a complete package because the only way to increase capacity of a wind system is to add an additional wind turbine or replace the one already installed.

Wind Hybrid Systems for Villages

Many remote villages throughout the world do not have electric power. In fact, about 30% of the world's population does not have electricity. The lack of electricity in rural villages is a major problem for many developing countries because people are moving to the cities, creating poor living conditions and food shortages. Governments have tried to use diesel generators to provide electricity to these rural villages, but poor maintenance due to a lack of trained repair people and no replacement parts have left most of these systems inoperable. The wind hybrid systems developed for telecommunications and remote homes in the developed countries of North America and Europe have now been developed further for remote villages. Oftentimes only a small amount of electricity needs to be provided to improve the quality of life greatly in a remote village where no one has had electricity.

Wind–diesel systems can be grouped into three categories based on the percentage of wind power that makes up the generating capacity. Penetration is the term used to describe the amount of wind power that contributes to the total generating capacity. There are two types of penetration, the first being installed penetration referring to the percentage of wind-generating capacity installed and connected to the system. The second is operating penetration and refers to the percentage of wind power actually being generated at the moment in relation to the total electricity being generated. Table 9.2 describes three different levels of penetration commonly used to describe the operations of electric grids that include wind power or a mix of power sources for electric generation [12]. Low penetration is what is found commonly in most utilities that allow wind power on their systems. For cases of low penetration, wind turbines are simply connected to the grid as described in the chapter on grid interconnected wind machines. The addition of wind power simply reduces the load on the diesel generators, but never comes close to totally unloading the diesel generators. Usually there is no supervisory control system that monitors the

Table 9.2 Levels of wind penetration on electric power grid and operating characteristics

Level	Average penetration	Operating characteristics
Low	<20%	Diesel always running
		Wind power reduces load on diesel
		Simple controls
Medium	20–60%	At least one diesel always running
		With high wind power, secondary loads energized to ensure sufficient diesel loading or wind power curtailed
		Simple supervisory controls required
High	>60%	Diesels may be shut off during high wind power periods
		Auxiliary components used to control voltage and frequency
		Requires sophisticated supervisory controls

entire generating system. The second level of penetration, mostly referred to as a medium level, is really the area between low and high. It has a fairly wide operating range and can range from 20 to 90% operating capacity with installed penetration being about 40 to 60%. In this mode, at least one diesel generator continues to run and operate with at least a 20 to 30% load while all the other generating capacity is supplied by wind power. A supervisory control system is required to turn on and off additional diesel generators and sometimes add additional loading to keep the voltage and frequency stable. This is usually a fairly simple type of supervisory control system. Finally, for the high penetration system, the wind power will be 100 to 300 or 400% larger than the system load or diesel generators. Other components have to be added to the system to help control the voltage and frequency, such as dump or auxiliary loads and energy storage devices. This system requires a sophisticated control system that manages all components in the generation system. In a high penetration system, there may be times that the system can operated without the diesel generators running. Controls for a high penetration system operating with AC electricity are much different than a system operating with DC electricity. Controlling electrical frequency requires a rotating electrical field to provide the driving frequency.

The National Renewable Energy Laboratory and the USDA–ARS teamed with Northern Power Systems to develop a wind–diesel hybrid system for remote Alaskan villages. The initial effort focused on three different villages of different sizes and different electrical demands, all operated by the Alaskan Village Electric Cooperative. The USDA–ARS constructed a small village wind–diesel system with three diesel generators, two wind turbines, a programmable village load, a dump load, and control system that could be modified easily. After several attempts, a workable system was developed and later implemented by Northern Power Systems in Alaska. A schematic drawing of the USDA system without storage is shown in Figure 9.16. This system is for a medium-level penetration system. An important lesson learned about this village hybrid system is that a lead diesel is needed to help with voltage and frequency control [13]. This generator may be supplying only a fraction of the load, but it is needed to provide frequency control and voltage control because the wind turbines will be following its frequency and voltage output. In addition to this lead diesel, a synchronous condenser is also needed to again help with frequency and voltage control. The synchronous condenser is another induction generator placed in the system without a power source. If a sudden drop in wind speed causing a loss of power or an increase in load happens, the

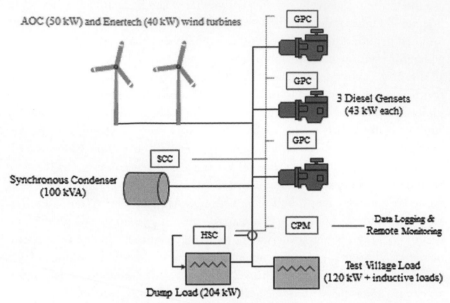

Figure 9.16 Schematic of a wind–diesel system for village power where wind penetration is in the medium level.

synchronous condenser will bridge the short change and allow the power sources to adjust to the new conditions. These changes in conditions occur so fast that normal control systems cannot react and make changes to keep a system stable. Medium penetration is the most common type of wind–diesel system installed today.

A wind–diesel system with battery storage for operation as a high penetration system is shown in Figure 9.17. The synchronous condenser was replaced by a rotary converter [14]. The rotary converter consisted of an AC motor/generator driving a DC motor/generator. When excess electric power was available from the wind–diesel generating system, power was sent to the rotary converter, which charged the battery bank. When the wind–diesel generating system was short of power, energy would flow from the batteries and be converted from DC to AC electricity. This system demonstrated that a wind–diesel high penetration system was feasible and could be developed. It was also determined that the control system must provide complete control to all components and that switches for the systems must be fast solid-state switches. Researchers concluded that such a system probably was not possible to construct prior to 2000 because the

Hybrid System With Storage

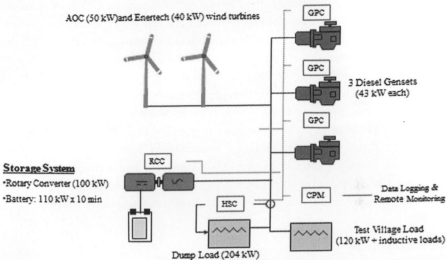

Figure 9.17 Schematic of a wind–diesel system for village power where wind penetration is in the high level.

electronics to perform the needed control function were not available. It is also recommended that an old diesel-generating system not be upgraded to a high penetration wind–diesel system because the diesel controls and engines are not equipped to meet the control demands. If you want to have

Figure 9.18 Wind turbines and diesel storage tanks at a wind–diesel generating plant in Selawik, Alaska.

a high penetration system, start from the beginning and build a complete new system. Retrofits should be reserved for medium penetration systems.

Figure 9.18 shows wind turbines and fuel storage tanks at the electric power plant in Selawik, Alaska. Selawik is a small village of about 800 residents and is located along a river in northwest Alaska. The power plant consumes about 120,000 gallons of diesel per year generating electricity. If it costs $5.00 a gallon for diesel delivered to the village, then they must spend approximately $600,000 just to purchase fuel. Wind turbines were selected to supply sufficient power to reduce the fuel consumption by 30 to 40%, which means that the village would save about $200,000 per year in reduced fuel purchases. The saved fuel gives a payback of under 10 years for the four wind turbines installed. Other Alaskan villages are having similar experiences, as well as many other villages around the Arctic Circle and resort islands in the tropics.

REFERENCES

[1] Golding EW. The Generation of Electricity by Wind Power. London: E. & F. N. Spon; 1955.
[2] Baker T Lindsay. A Field Guide to American Windmills. University of Oklahoma Press; 1985.
[3] Aermotor Wind Company, http://aermotorwindmill.com.
[4] Righter RW. Wind Energy in America, a History. University of Oklahoma Press; 1996.
[5] World Wind Energy Association (WWEA). Small Wind World Report. New Energy, Husum 2012. March 2012.
[6] Clark RN. Performance comparison of two multibladed windmills. SED vol. 12, 11th ASME Wind Energy Symposium, ASME 1992. p. 147.
[7] Ling S, Nelson V, Clark R, Vick B. Field testing a smart controller for wind–electric water pumping systems. A collection of 2000 ASME wind energy symp technical papers at the 38th AIAA Aerospace Sciences Meeting and Exhibit. Reno, NV: AIAA Paper No. 2000-0055, 2000. pp. 339–45.
[8] Vick BD, Clark RN. Performance and economic comparison of a mechanical windmill to a wind-electric water pumping system. ASAE Paper No. 97–4001, 1997.
[9] Vick BD, Clark RN, Ling J, Ling S. Remote solar, wind and hybrid solar/wind energy system for purifying water. Journal of Solar Energy Engineering 2003;125:107–10.
[10] Gipe P. Wind Power, Renewable Energy for Home, Farm, and Business. White River Junction, VT: Chelsea Green Publishing Co; 2004.
[11] Perez, R. Publisher. Home Power magazine. http://homepower.com.
[12] Nelson V. Wind Energy, Renewable Energy and the Environment. Boca Raton, FL: CRC Press; 2009.
[13] Eggleston E, Clark RN. USDA wind hybrid research laboratory controls development. Proc 12th Annual International Wind-Diesel Workshop, Prince Edward Island, Canada: Paper #4; 1998.
[14] Clark RN, Eggleston E. Wind hybrid operation with pre-commercial controls. Proc. 14th Annual International Wind-Diesel Workshop, Prince Edward Island, Canada: Paper #9; 2000.

CHAPTER TEN

Economic Considerations
Predicting the Economic Reality of an Installation

With contributions from Trudy Forsyth and Frank Oteri

Everyone wants to know, "What's it going to cost?" and "When will it pay for itself?" These are simple questions, but their answers are hard for small wind systems because so many of them depend on wind resource variability, installed turbine costs, and competing electricity sources. Small systems are installed at homes to reduce the amount of purchased electricity, so their economics depends on the retail cost of the electricity they are displacing. This is in contrast to the machine installed at a remote cabin where the energy produced provides the only power available. How do you determine the value of power that was nonexistent before installation of the wind system?

The economics for small wind machines is much different from the economics for large wind machines when installed in wind farms, where the electricity provides wholesale value. The energy produced by small machines is consumed at the site and by the owner. Therefore, its value is based on what it is worth as it is used. It could be replacing very expensive energy if it is in a remote location where electricity is produced by a diesel generator or similar device, or it could be replacing rather inexpensive electricity if connected to a strong electric grid. However, the value of electricity produced for sale to a large electrical grid is based on the sales contract and the ability of the machine to produce power. Uncertainty in economic prediction resides in the estimation of the electricity produced; therefore, the prediction must be as close to actual conditions as possible.

The two most important things to consider in determining the economics of a wind machine are the machine's installed cost and its annual energy production. There are other costs to consider, but they do not impact the profitability of the machine as much as these two do.

Small Wind
ISBN 978-0-12-385999-0

THE TRUE COST OF WIND MACHINES

In addition to the actual cost of the machine, the true cost incorporates several other factors. It includes the tower, the foundation, the wiring, and any electrical modifications made in the house to accommodate the interconnection hardware. Associated with the installation may be permit fees, interconnection fees, and other charges related to getting the turbine installed and operating. This cost may be offset or reduced by investment tax credits or other grants that may be available. In the United States, a 30% federal investment tax credit is available for machines less than or equal to 100 kW installed before the end of 2016. Be sure to check for state and local programs offering tax or other credits for installing wind systems. The web site www.dsireusa.org maintains a listing of current state and utility credits and grants. Do not include loans as a reduced installation cost because you must repay them, even though they may be at low and reduced interest rates.

A second type of cost that must be included is the interest on any loans related to the purchase and installation of the wind machine. This interest is a cost that must be applied against the machine. Also, you may want to charge some lost interest against the machine if you paid cash for it because of lost revenue from monies not invested.

Be sure to include some annual cost for maintenance. About the only operational cost from scheduled maintenance is that for oil and grease to service the machine's gearbox and bearings and for your time. Typically, machines are inspected thoroughly once a year and the gearbox oil is changed and bearings are lubricated; there is also a turbine inspection per the manufacturer's recommendations. Several studies have been done on the cost of maintenance for wind machines and have shown that most systems range between $0.01 to $0.015 per kWh, with machines with gearboxes and hydraulic brakes costing the most and machines with only bearings costing the least. These maintenance costs do not include costly replacement parts such as generators or blades, but they do include the replacement of more routine components. Estimates of maintenance costs are getting better and lower because machines are better built and scheduled inspections are better at finding potential problems before they become significant repairs.

Insurance for the machine is a cost that is often not included in an economic analysis. If you borrowed money to purchase the machine, the lender will most likely require that you carry some insurance against damage to it. Two types of insurance you may want are coverage for repairs or replacement and accident coverage in case the wind machine damages someone's property or causes someone injury. As more and more small wind systems are installed, larger insurance companies are establishing riders or attachments to current homeowner policies to cover wind systems. Others are still requiring that a separate policy be written. It takes considerable time and effort to seek out a reasonable insurance policy that provides the desired coverage and is affordable. Just make sure these costs are included in the economic analysis.

The final item to be considered when estimating the cost of a wind system is taxes to be paid or in some cases deducted. Wind turbines add value to the property; therefore, they increase the property assessment for tax purposes. Many local governments offer tax exemptions for wind systems installed at residences, but charge a property tax for businesses. Make sure to check the tax codes at the location where you are installing a wind machine. The property tax issue is still being modified in many locations. If money is borrowed to purchase the wind system, the interest paid on the loan may be tax deductible, thus decreasing the amount of income tax paid. The amount of this saving depends on other factors such as tax bracket and other taxable income.

INCOME FROM WIND MACHINES

A wind turbine produces income in the form of displaced energy, usually as electricity for direct use and as a surplus, again in the form of electricity that is sold. In the case of water pumping or other stand–alone applications, the wind machine may displace the purchase of another energy source and its fuel—that is, the purchase of a diesel engine and the fuel it consumes over some projected lifetime. Projected income can be hard to estimate and depends on the application and expected use. Annual energy production can be estimated fairly closely if the average wind speed and rotor area of the machine are known. As more machines reach certification, the accuracy of annual energy projections should improve. Still, a major issue is the relative percentages of the annual electricity to be valued at the full retail rate and at the payback rate. Data analyzed in Chapter 3 indicate that even if the turbine supplies only 25% of average use, there are significant times of electrical feedback to the utility accounting for 15 to 20% of the energy generated. Thus, a realistic number might be 80% to 85% to be valued at the retail rate and the remaining 15% to 20% to be valued at a lower payback rate. If a purchase agreement has not been completed, it may be almost impossible to determine a value of the electricity produced and the income from the wind turbine.

Most economists include an escalation factor for the cost of displaced electrical energy. The rate of price increases for electrical energy varies so much from utility to utility that it is hard to suggest a number. Utilities that have built generating plants fueled by nuclear and oil have shown large price increases since the early 1990s, whereas increases for utilities that have built new plants fueled by natural gas and coal are lower. Considering a conservative number for escalation of the price of electricity makes the payback longer, but it will not give you an unrealistic number as your projection.

Simple Payback

The time at which savings or earnings equal the cost of the wind machine is the payback time. For the system to be economically feasible, its overall earnings should exceed the cost within a time period that is less than the machine's lifetime. Because wind turbines have a high initial cost, their payback is usually higher than that for other items purchased. A payback of approximately 10 to 15 years is not uncommon. Paybacks exceeding 20 years should be viewed with caution because components begin to wear out and may require significant repairs. The simple payback can be determined by dividing the total installed cost by the cost that is displaced each year through saved energy, less any operation and maintenance (O&M) cost:

$$SP = IC/[AEP * \$/kWh - AOM] \qquad [10.1]$$

where SP is the simple payback in years, IC is the total installed cost in dollars (or country currency), AEP is the anticipated annual energy production in kWh/year, $/kWh is the price paid for the electricity displaced, and AOM is the annual O&M cost in dollars (or country currency).

Example 10.1

A small wind turbine is purchased for $15,500 and produces 5,300 kWh per year. The cost of electricity is $0.16/kWh and the O&M cost is estimated at $75. The result is

$$SP = 15,500/5300 * 0.16 - 75$$

$$SP = 15,500/773 = 20 \text{ years}$$

If you want to consider lost income from interest not earned on monies invested in the wind system, or if you borrowed money for the purchase, the interest is subtracted from the income provided. Equation 10.1 then changes to the following:

$$SP = IC/[AEP * \$/kWh - IC * FCR - AOM] \qquad [10.2]$$

where FCR is the fixed charge rate per year. FCR represents interest lost or interest on money borrowed. If part of the installed cost was borrowed, you use the borrowed amount rather than the full installed cost to calculate the interest loss.

Example 10.2

Using the same wind turbine from Example 10.1, but considering that all installed costs were paid for by money borrowed at an interest rate of 4%, the result is

$$SP = 15,500/[5300 * 0.16 - (15,500 * 0.04) - 75]$$

$$SP = 15,500/153 = 101 \text{ years}$$

The addition of lost interest or interest on borrowed money increases the payback time by a factor of 5. If the interest rate drops to 3%, the payback period drops to 50 years, indicating the significance of the interest rate.

Several assumptions are made for these simple payback calculations. The most important is that the price of displaced electricity remains the same throughout the total time period and there is no allowance for inflation. Another, more sophisticated analysis, to be discussed later in this chapter, may reduce the payback period. This simple method gives an approximation of what might be expected from a wind system.

Cost of Energy

Another parameter often used to determine the potential economic value of a wind system is the levelized cost of electricity (LCOE) from it, which is compared to the cost of other electrical sources. The LCOE calculation gives a more accurate representation compared to simple payback in part because it considers all of the cost elements required to install the turbine. It does not account for any change in electricity rates over time, leaving the estimate on the conservative side. Also, the cost of electricity is often used to determine the relative efficiency of different wind turbines or wind turbine types.

Again, the cost of electricity is dependent on the installed cost and the annual electrical production.

$$LCOE = (IC * FCR + AOM)/AEP \qquad [10.3]$$

where LCOE is the cost of electricity in \$/kWh, IC is the installed cost in dollars (or country currency), FCR is the fixed charge rate or the percentage interest on borrowed money, AOM is the annual O&M in dollars, and AEP is the annual energy production in kWh. The fixed charge rate is included to account for the interest paid on borrowed money or for lost revenue from monies not invested. Annual O&M includes replacement of some components over the lifetime of the wind turbine.

Example 10.3

Again we use the same wind turbine from Examples 10.1 and 10.2, where the installed cost of the turbine was \$15,500, the annual electrical production was 5,300 kWh, and the annual O&M was \$75. An FCR of 4% is also used. The cost of electricity is found to be

$$COE = (15,500 * 0.04 + 75)/5300$$

$$COE = 695/5300 = \$0.131/kWh$$

The COE calculation gives the prospective buyer a comparison of the cost of electricity produced, which can then be compared to the cost of competing electrical sources. This cost should be the same for the next 20 to 25 years, compared to electrical sources that will have fuel cost and regulatory increases.

Figure 10.1 is a COE sensitivity chart that can be developed for any installation [1]. Installed cost, annual energy production, and annual O&M are increased and decreased by percentage increments. The data show the effects of a change in any of the parameters from −20% to +20%. The chart illustrates the importance of a site that produces the most energy and why it is important to place the wind turbine in a free flow of wind. At the same time, the chart shows that installed cost also influences COE. Even though annual O&M costs are important, they do not influence COE as much as annual energy production or installed cost do.

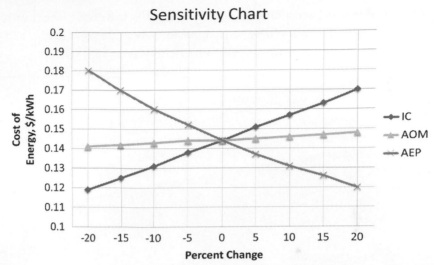

Figure 10.1 COE sensitivity chart showing the effects of changes in installed cost, annual energy production, and annual O&M cost on the cost of energy produced by the wind machine.

CASH FLOW ANALYSIS

Although simple payback and LCOE are common, often discussed economics terms, many wind experts suggest that the only true economic analysis that needs to be conducted is a cash flow analysis. With cash flow analysis, additional factors that influence the effectiveness of a project can be included. Probably the most important factor in such an analysis is potential changes in the electrical rate charged by the utility. Including a utility inflation factor clearly shows the impact that increased electric prices have over the 20- or 30-year life span of a wind system. The impact of paying a loan can also be evaluated in a cash flow analysis.

The following inputs are needed to do a cash flow analysis. Some of the information will be known; some of it will have to be estimated.

Total installed cost ($)   Federal tax credit (%)
Annual energy output (kWh)   Electricity cost from utility ($/kWh)
O&M cost ($/kWh)   Electricity inflation rate (%)
O&M inflation rate (%)   Amount of loan ($)
Insurance premium ($)   Down payment ($)
Insurance premium inflation rate (%)   Interest rate (%)
State rebate (% or $)   Loan term (years)
State tax credit (%)

After you have gathered the information needed, a cash flow table can be constructed. The table should be carried out for at least 25 years to give a full picture of the possibilities for the machine being considered.

●●●

Example 10.4

The information used in this example is shown in Table 10.1 and is for a home or small business wind machine. It assumes that all energy produced will be used to offset the full price of electricity purchased from the utility. A column must be added to account for any electricity sold at a rate below full retail electricity rates. Inputs and assumptions are as follows:

- Installed cost: $45,000
- Annual energy production: 22,000 kWh

Table 10.1 Annual cash flow

Year	Sale of energy	O&M costs	Insurance	Loan payments	Annual cash flow	Total cash flow
0					−15000	−15,000
1	3,300	0	−450	−4,800	−1,950	−16,950
2	3,366	0	−464	−4,800	−1,898	−18,848
3	3,433	0	−477	−4,800	−1,844	−20,692
4	3,502	0	−492	−4,800	−1,790	−22,481
5	3,572	−225	−506	−4,800	−1,959	−24,441
6	3,643	−232	−522	−4,800	−1,910	−26,351
7	3,716	−239	−537	−4,800	−1,860	−28,210
8	3,791	−246	−553	−4,800	−1,809	−30,019
9	3,866	−253	−570	0	3,043	−26,976
10	3,944	−261	−587	0	3,096	−23,880
11	4,023	−269	−605	0	3,149	−20,731
12	4,103	−277	−623	0	3,204	−17,527
13	4,185	−285	−642	0	3,259	−14,269
14	4,269	−294	−661	0	3,314	−10,954
15	4,354	−302	−681	0	3,371	−7,583
16	4,441	−311	−701	0	3,429	−4,154
17	4,530	−321	−722	0	3,487	−667
18	4,621	−330	−744	0	3,547	2,880
19	4,713	−340	−766	0	3,607	6,486
20	4,807	−351	−789	0	3,668	10,154
21	4,904	−361	−813	0	3,730	13,884
22	5,002	−372	−837	0	3,793	17,677
23	5,102	−383	−862	0	3,856	21,533
24	5,204	−395	−888	0	3,921	25,454
25	5,308	−406	−915	0	3,987	29,441

- Cost of electricity: $0.15/kWh with 2% inflation
- O&M cost: $0.005/kWh with 3% inflation (5 year warranty)
- Insurance: 1% of IC with 3% inflation
- Loan: $30,000 with $15,000 down payment

A few manufacturers provide a sample cash flow table on their web sites that you can download and enter your own numbers to evaluate a potential purchase. One of the best is at Bergey.com [2]. Gipe has example tables in his book, *Wind Power, Renewable Energy for Home, Farm, and Business* [3].

The cash flow analysis can also be presented in graphical form, which clearly shows the time when the cash flow changes from negative to positive; the point where the cash flow crosses the zero line is the anticipated payback period. The data from Table 10.1 are shown plotted in Figure 10.2. The cash flow debt continues to grow until the loan is paid, and then it improves until all of the investment has been satisfied in year 17.

Other types of economic analysis are sometimes used, but they are mostly reserved for larger business and industrial projects. Determining present worth and determining levelized costs are two methods that include more detailed calculations of changes in interest rates and inflation rates. However, most purchasers of small wind systems will not want to include the extra expense and time required to refine the numbers for their economic analysis. One important aspect of an economic analysis is that it

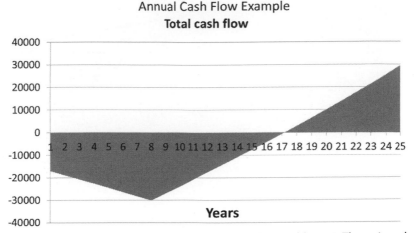

Figure 10.2 Results of a cash flow analysis using data from Table 10.1. The point where the cash flow crosses zero is the payback time.

clearly shows where changes or improvements can be made in a project to ensure economic success—for example, a taller tower to increase annual energy production with little additional installed cost. Results may be a four to five year reduction in payback time.

REFERENCES

[1] Vosper FC, Clark RN. Energy production and performance of a wind-driven induction generator. Trans ASABE 1985;28(2):617–21.
[2] Bergey Windpower web site, www.bergey.com.
[3] Gipe P. Wind Power, Renewable Energy for Home, Farm, and Business. White River Junction, VT: Chelsea Green Publishing Co; 2004.

CHAPTER ELEVEN

Operation and Maintenance
A Guide to Long-Term Operation and Maintenance Issues

There has been a wind industry for centuries, but except for a few innovative changes that occurred over time, nothing has significantly impacted the industry so much as the availability of power electronics and computer technology. Thus, the wind industry has changed significantly since the mid-1970s. Electronics have not only impacted the techniques for generating electricity, but have also greatly impacted the ability to operate and control a wind rotor operating in a harsh windy environment. Speed and power control of Dutch-design wind machines was accomplished by stopping the machine and either adding or decreasing the amount of cloth sails covering the wooden–slatted rotor blades. Adjusting to a different wind speed often required two to three hours. Inclusion of an offset, spring-loaded tail vane allowed for the American windmill to adjust to changing wind speed conditions rapidly and even to almost stop the rotor wheel in very high winds. All these controls were either by hand or mechanical and would not work for machines generating electricity where the generator offered little resistance if the electric field was lost.

Many operating issues are now being addressed with small single-board computers called microprocessors and by power electronics. This is an area where small machines have benefited from large machine development because manufacturers of large machines could afford to invest in the development of new controllers and sensors designed for wind machines. The evolution in wind turbine controllers exactly parallels the evolution of computers, going from large desk-sized units to the small hand–held technology of today. Early controllers could monitor only a few parameters, but modern controllers can monitor hundreds of them and process their inputs and make adjustments to maximize operation.

Manufacturers have suffered in the past because turbines did not operate well, experienced many failures, and often did not meet operational expectations. Because of the failures of the past, some manufacturers tend to overexaggerate the capabilities of their machines and indicate that they will run for years without maintenance. Remember that wind turbines are machines and that machines need service and maintenance. Automobiles are much better today than they were in the early 1990s, but they still require maintenance. The same is true with wind turbines; however, the level and types of maintenance vary depending on the components within the wind turbine. It is not something that is placed on a tower and forgotten—it needs to be watched. The watching is much easier now with the new smart controllers that can provide almost instantaneous feedback and data from the wind turbine. So watching the machine may involve checking a file on the computer or a display on the kitchen wall. An example of a turbine status report is shown in Figure 11.1; it was provided by a friend of the author who has a Skystream 3.7 [1]. The diagram shows the turbine status in the upper left-hand corner and the communication status in the lower left-hand corner. In the center is a graph showing the daily energy production for the last 30 days. Current power output and rotor speed are shown along the bottom. Finally, a table shows the total kWh for the month, average daily

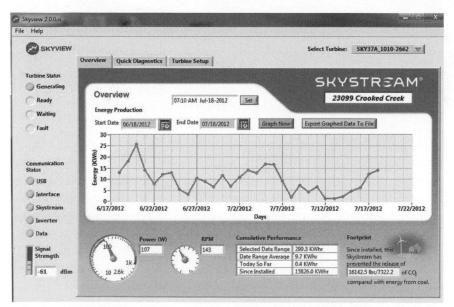

Figure 11.1 Screen shot of wind turbine–tracking software that allows owners and operators to view machine performance. The program is Skyview by Southwest Wind Power [1].

Figure 11.2 Wireless interface that communicates with the wind turbine and transfers data to a computer.

energy production during the month, total for the day, and total since installed. A small wireless box communicates with the wind turbine to download these data to a computer for display (Figure 11.2).

Another example of a turbine status is shown in Figure 11.3, taken from a public web site maintained by Northern Power Systems [2]. The diagram shows the wind turbine actually turning as if it were producing power and

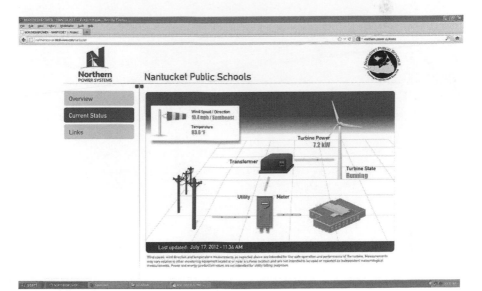

Figure 11.3 Screen shot of the Public View page of the Northern Power Systems web site showing the current status of a turbine at a school [2].

the flow of electricity to the load or to the utility. Figure 11.3 also shows wind speed and temperature. With a quick look at this screen, one can determine if the machine is operating and, after some experience, whether it is operating near normal by comparing turbine power and wind speed. The software system that provides this type of wind turbine monitoring is called a supervisory control and data acquisition (SCADA) system. These systems were developed to monitor wind turbines in larger wind farms and, when first developed, their interconnections were all hardwired. Now, with wireless technology, these systems are available for small individual systems installed where a wireless network can be established.

Some larger wind turbines, especially those located in remote areas, may include much more information from their SCADA system, similar to the one shown in Figure 11.4. This SCADA includes not only information about the wind turbine operation, but will show items that need attention from a maintenance crew. It also shows some faults that would cause an emergency shutdown. This form of SCADA may be used more in research

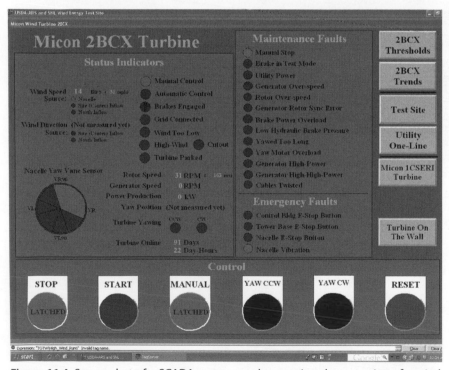

Figure 11.4 Screen shot of a SCADA system used to monitor the operation of a wind turbine.

and development, but it provides an idea of what the capabilities are of these new control systems. Some manufacturers have included software in their controllers that allows new operating programs to be downloaded into an existing machine without having to climb the tower or visit the site. Servicing can be carried out remotely via wireless connections. This new technology has improved the capability to monitor and watch small wind machines greatly.

INSPECTIONS

All wind turbines need to be inspected periodically to observe if they are operating properly. As one installer noted, it should be more than checking to see if the turbine is on the tower and spinning. The owner or operator should spend a few hours watching and listening to a new machine. It is important to note how it tracks or follows the wind and to note if it acts slightly different when the wind is from a particular direction. There may be some obstruction that causes the turbine to osculate more when the wind comes from that direction. These observations help you as the operator recognize when the machine may not be acting normally. Another important thing to do is to listen to the machine running. Often a change in the noise level will be the first sign of a potential problem. Simple issues can often be observed by listening to the turbine operate. Once, a machine started making a louder swishing noise as it rotated; upon inspection it was found that the leading edge tape had come loose on one blade, causing that blade to create more resistance and thus more noise. Replacing the tape solved the problem and allowed the turbine to regain its potential output.

Watching and listening to the turbine for a few minutes every week or more often can often indicate potential problems before major repair issues occur. If something is suspected, then the monitoring system can be checked or a more thorough inspection may need to be done. Binoculars are a helpful investment for small wind turbine owners or operators because they offer a way to get a close-up view of the turbine without climbing a tower or laying a tower down for some visual inspections. They are helpful in locating loose objects that may be heard knocking or flapping in the wind. The important thing is to make consistent and frequent visual and listening inspections.

Once a year, it is important to do a thorough inspection of the unit. This means that a trained repairman should climb the tower and inspect all aspects of the machine (Figure 11.5). All bolts, including tower bolts, should be

Figure 11.5 Wind technicians inspecting a 10-kW wind turbine.

checked for tightness and wear. The yaw bearing should be checked and, if possible, lubricated. All other bearings need to be checked and lubricated as well. Units equipped with gearboxes should have the fluid level checked and, if indicated by the owner's manual, the fluid changed. All brakes should be checked for excessive wear and pads changed if necessary. The electrical wiring should also be checked and should include the slip rings if they are part of the system. This should be a thorough inspection of the system to make sure all is operating properly and that no part shows excessive wear. A good yearly inspection will help ensure continued operation and minimize potential problems.

COMPONENT REPAIRS

Visiting with any experienced installer who does maintenance and installations will provide you with information about what components to watch or those to be concerned about. There have been very few written reports about component repairs for small machines because no one wants to reveal their problems. However, one report, written by the U.S. Department of Agriculture (USDA)–Agricultural Research Service, describes the repair and maintenance of a 50 kW wind turbine after 20 years of operation [3]. It should be remembered that the machine studied was a prototype machine that experienced some design changes throughout the reported period. Regardless, this report gives some suggestions about what components should be watched and need the most maintenance and repairs. Figure 11.6 shows two pie charts with the available runtime in one and the components that needed repairs in the other. First, it should be noted that

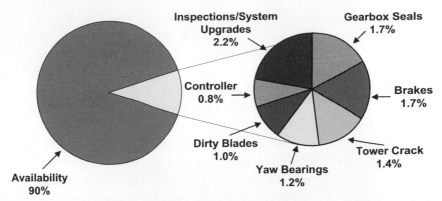

Figure 11.6 Summary of components requiring repair and maintenance during 20 years of operation of a 50 kW wind turbine. *(Data reported by Clark [3].)*

this machine was available for operation 90% of the hours for the 20–year period. The machine actually produced power more than 100,000 hours and provided approximately 1.5 MWh of electricity.

Two significant repairs that required the turbine to be taken down were replacement of an upper tower leg and replacement of the yaw bearing. In both cases, the nacelle had to be removed and the parts ordered from the factory before repairs could be done. Each of these repairs required about 2,500 hours or 3.5 months each to complete, thus consuming about one-third of the downtime. Another third of the time was used in maintaining the gearbox and brakes. Gearboxes have continued to be a significant item of maintenance on all wind machines that use them, which is why small wind turbine designers have opted to use permanent magnet alternators without gearboxes. These designs just avoid potential issues with gearboxes. The biggest problem is maintaining the seals around the shafts, especially the input or low–speed shaft. For the turbine reported in the study, the seals had to be replaced about every five years. Again this repair required that the turbine be removed from the tower because the blades and hub had to be removed to slip a sleeve over the shaft. This repair is much like replacing the seals on an engine crankshaft.

Brake pads wear out and have to be replaced. The frequency of brake pad wear was dependent on the sequence and number of stops and the number of emergency braking incidents. Turbines that slow the rotor electrically before applying the disk brake experience less wear than those that use the brake to supply all the stopping power. Brake pads on the machine reported had to be replaced about every three years. Again, this is why some small machines do not use a mechanical brake with pads, but rely on a resistance brake. Caution should be exercised when applying a resistance brake to a machine with a permanent magnet alternator because the switch actually shorts the three electrical leads together to create the resistance. If this is done at a wind speed above 8 m/s, the output may overpower the resistance and cause the windings to be burnt beyond repair. Many early users of these braking systems learned this lesson the hard way by burning out their alternator windings. The operator's manual should give detailed instructions on how to use the braking system.

The last third of downtime described in the USDA report is identified with inspections as described earlier and system upgrades, controller issues, and dirty blades. Small machines that are purchased that have completed certification testing should not be having issues with system upgrades unless it is something similar to an automobile recall where all units that have been produced are modified. The greatly improved electronics being used in presently sold controllers should also limit the amount of downtime. One exception to the electronics issue is improper grounding of the wind turbine, as lightning can cause significant damage to the electronics. Make sure that the machine is grounded properly; checking the grounding should be included in any annual inspection.

The final area of downtime is dirty blades. New wind turbines use entirely different airfoils than do machines manufactured in the 1980s; therefore, their susceptibility to dirt and bug buildup has been designed out of the blades. However, during the summer and fall of 2011, several wind turbine owners did report some lower power production that appeared to be due to dirty blades. This was after a period of two to three months without rainfall. Normally, there is sufficient rainfall to clean the blades in most years.

Most dealers and installers will provide some type of inspection and routine maintenance agreement that will take effect after the manufacturer's warranty expires. They will have experience with operation of the machine and will know the components that will be potential problems based on past experience. Repairs should be performed by qualified personnel who have had training and experience in dealing with the machine. Also, they can guide you on how to conduct weekly or monthly observations.

EMERGENCY SITUATIONS

As stated before, small wind turbines are machines and they do fail just like your car, home heater/air conditioner, or any other machine around the house. Some fundamental precautions can be taken to help prevent more damage to the wind turbine or the creation of a hazardous safety concern for nearby property and people. First and foremost, learn where the electrical disconnect and emergency stop switches are located and make sure the path to them is never blocked. Events such as icing of blades or power outages in high winds may necessitate stopping the turbine. Other weather–related events may suggest that the turbine be stopped for a few hours rather than risk permanent damage to the turbine. All these events can be minimized by simply knowing how to stop the turbine and disconnect it from the electric grid.

If you have been observing the turbine on a routine basis, it is fairly easy to determine if it is not working properly. If this is the case, stop the machine if you can and call for an experienced repair person or team to come check the turbine. Do not try to climb a tower with the machine turning. This is never wise, even for an experienced repair person. If the turbine cannot be stopped, keep all people away from the turbine and upwind if possible. Experience has shown that most falling pieces from wind turbines fall within a distance from the tower base that is less than the height of the tower. Blade pieces have been thrown further, but those pieces are usually small and normally don't cause much damage. They always go to the downwind side of the turbine, but may fall within a 100° arc. This is why it is important to stop the machine if possible with the braking system. Once the rotor is stopped, the rotor can be secured by tying the blade to the tower. Again, an experienced tower climber with appropriate climbing safety gear should perform this task, as shown in Figure 11.7.

Figure 11.7 Wind technician tying off a rotor blade before beginning repairs on a wind turbine.

Always create a safe working environment before working on any component of the wind system. If someone is to work on the wind turbine, make sure that all external electrical power to the turbine is turned off before approaching the tower. Also, have provisions so that someone cannot turn on the system while someone else is on the tower working. The person on the tower should have assurance that no one can switch anything on before he is ready. The first thing to do when climbing the tower is to secure the rotor so that it cannot rotate or yaw. A sudden change in wind direction can cause the turbine to turn, possibly pinning the repair person against the tower or breaking his safety strap. The same is true for a rotating blade, as a 3-bladed rotor turning at 2 rpm means that a blade is coming at you every 10 seconds. Those are good odds for being struck. Of course, the larger the machine, the more dangerous it is working on the tower if it does not have a work platform. After all repairs are made and everyone is on the ground in a safe place, then the turbine can be restarted.

REFERENCES

[1] Neal B. Personal communication, July 17, 2012.
[2] Northern Power Systems web site, www.northernpower.com (Public View link).
[3] Clark RN. Performance and maintenance experiences with a wind turbine during 20 years of operation. Proc. of AWEA Global Windpower 2004, Chicago, IL: CDROM 2004.

CHAPTER TWELVE

Distributed Wind Systems
Promoting Growth of Small Wind Systems

With contributions from Trudy Forsyth and Frank Oteri

"Distributed wind energy systems provide clean, renewable power for on–site use and help relieve pressure on the power grid while providing jobs and contributing to energy security for homes, farms, schools, factories, private and public facilities, distribution utilities, and remote locations. Distributed wind systems generally provide electricity on the retail side of the electric meter without the need of additional transmission lines" [1]. Distributed wind is commonly referred to as small wind, medium wind, and community wind when using wind turbines at locations to offset all or a portion of onsite electricity consumption. Actually, any installation that allows for the generated power to be consumed at or near the point of production can be classified as distributed wind. The term "distributed wind" is much more common now because in 2010, a group of manufacturers, distributors, project developers, dealers, installers, and advocates formed the Distributed Wind Energy Association (DWEA). The primary mission of DWEA is to promote and foster all aspects of the American distributed wind energy industry. The fundamental goals of DWEA are to [1]

- Develop a federal, state, and local policy environment that supports the responsible expansion of distributed wind energy
- Reduce or eliminate unwarranted barriers to the use of distributed wind energy
- Provide a unified voice for all members and sectors of the distributed wind industry
- Develop and promote industry "best practices" policies and standards that will foster the safe and effective installation and operation of distributed wind systems
- Participate in public and consumer education

It is important to note that distributed wind does not identify itself with a particular size of wind turbine, but rather on the point of entry of the locally generated power into the electrical grid or power system. Distributed

wind contrasts itself from wind farms or wind plants, which provide electrical power on the wholesale side of the meter and must have substations
and larger transmission facilities to move their power to a point of
consumption. The author sees the wind industry grouping itself into three
categories, each with its own list of issues and concerns: wind farms
providing wholesale power, distributed wind providing onsite power at the
retail level, and offshore wind providing wholesale power much like wind
farms, but with very different installation and operational issues.

This chapter discusses the three different elements to distributed wind
systems and gives examples of how they operate and contribute to the
concept of distributed wind. First, several installations of wind systems
provide onsite electrical energy, but are larger than the definition of small
wind that is in the design and safety standards written by international
experts through the International Electrotechnical Commission resulting
in "Design Requirements for Small Wind Turbines" IEC-61400-2 [2].
Many of these installations fall in a category that some people call
"midsized turbines" (101 kW to 1 MW). The Interstate Turbine Advisory
Council recently defined medium wind as turbines with a swept rotor area
between 200 and 1,000 square meters. A number of these turbines left
over from past wind development activities are being refurbished and
installed in distributed situations. Also, several manufacturers have
returned to making units in this size range because they are easier to
transport and install in many remote locations where they often deliver
electricity at competitive prices.

Another unique area of distributed wind systems is referred to as
"community wind" because of the method that the turbine is owned and
how the revenue is divided. As the name suggests, the turbine is owned by
a number of investors who share the cost of ownership equally. Each
investor is entitled to share in the power production as credit toward the
power used on his nearby property or share in the financial benefits
generated by the wind turbine. This concept of ownership was first used in
Denmark in the 1980s, but has been slow to get going in the United
States. It has only been since 2005 that this type of ownership and
distribution of the production has been allowed in the United States [3].
Machines installed may or may not fit the small machine criterion of being
less than 100 kW.

Finally, the third component of distributed wind systems is small
machines less than 100 kW. Assuming that the main purpose of many small
wind systems is to provide electrical power for a family, it is noteworthy to

consider the yearly electrical use and what size turbine is needed for households in different parts of the world. In America, an average family uses about 11,500 kWh per year. Typically a 10 kW wind turbine would supply the needs for this American family. As a comparison, a European household demands a 4 kW wind turbine and an average Chinese household demands a 1 kW wind turbine [4]. Prior to 2008, about 80% of all small wind machines sold in the United States were for off-grid applications. Between 2008 and 2012, that number had decreased to 60%, meaning that more small machines were being used for on-grid applications. However, in 2011, less than 5% of small machines sold were larger than 10 kW. Both the United States and Europe have seen a significant increase in grid-connected machines between 1 and 10 kW [5].

MIDSIZED TURBINES USED IN DISTRIBUTED WIND SYSTEMS

Many wind turbines larger than 100 kW are being used in locations where power is consumed before it reaches the utility grid. Some of the first of these were installed at schools and community colleges. The Spirit Lake, Iowa, school district installed a 250 kW wind turbine in 1993; it was so successful that the district installed a second turbine in 2001 (Figure 12.1). The 250 kW Windworld turbine powers the elementary school and cost $239,500 to install. Partial funding for this turbine came from a $119,000 grant provided by the DOE. The remaining project cost was funded through a low-interest loan from the Energy Council of Iowa, Department of Natural Resources. The average annual savings from the turbine during the period 1993–1997 was $21,673 (Table 12.1). In 1998, the district made the final loan payment—3.5 years ahead of schedule [6].

In 2001, the district installed a 750 kW NEG Micon turbine for a total installed cost of $780,000. A $250,000 loan with no-interest financing was provided by the Iowa Energy Center's Alternative Energy Revolving Loan Program. The Iowa Department of Natural Resources assisted with a $580,000 loan at an interest rate of 5.1% per year. During its first five years of operation, the NEG Micon averaged approximately 83% of its projected output with an annual production of approximately $120,000. This exceeded the district's yearly loan payment of $93,000. From 2001 through 2011, the 250 kW Windworld and the 750 kW NEG Micon averaged nearly $137,000 in annual production, offsetting approximately 46% of the

Figure 12.1 Wind turbines at Spirit Lake, Iowa, public schools. Example of midsized turbines installed as distributed wind systems (NREL).

Table 12.1 Spirit Lake School District savings from onsite wind turbines (second turbine added in 2001)

Year	Savings ($)
1993	12,243
1994	24,978
1995	25,007
1996	25,951
1997	20,185
1998	19,853
1999	23,969
2000	21,650
2001	46,123
2002	137,798
2003	136,406
2004	150,420
2005	156,474
2006	129,987
2007	146,018
2008	132,776
2009	128,377
2010	163,515

district's electricity needs. The school used a combination of grants and loans to purchase the wind turbines and reported that they were completely paid off in 2010. Now all the savings go into a general fund for student instruction.

In talking with several school superintendents, the author discovered that most schools have to use local tax money to pay for utilities and can use state and federal funds only for books, supplies, and teachers' salaries. When utility costs increase, the only way to pay them is to raise local property taxes. For this reason, local school boards are looking for ways other than increased taxes to cover potential increases in utility costs. Purchasing a wind system locks in a fixed price for electricity over a 20-year period and provides some level of protection against large increases in electric rates. Many schools are seeing success with wind.

Midsized wind turbines are being installed at farms and businesses, but not as many as at schools and other public facilities. The economics of the installation depend on the availability of state and federal incentives, local net metering and interconnection policies and the zoning and permitting requirements. Many states cap their net metering at something less than 100 kW, thus not making it available to a midsized system. Also, almost all businesses have at least a two-component electrical rate. The actual energy consumed is one component and a second component is a demand charge. The demand charge is a flat fee charge based on maximum usage during a 15-minute period during the last year or six months. The exact method of determining the demand charge varies among the utilities and cooperatives that provide the retail electrical service. Also, many businesses require a shorter payback period for capital investments than the 7 to 10 years typical with individual wind turbine ownership. This is why many of the business installations are at family-owned businesses rather than at larger corporate-type businesses, which focus on short-term profits.

Some larger businesses with large electrical requirements have turned to wind power to offset their electric bills. One example is a gasoline refinery operated by Valero Energy Corporation where they installed 33 wind turbines to offset some of their $1.4 million a month electric bill [7]. Wind turbines were installed around the plant and were connected to the plant's local electric grid (Figure 12.2). The anticipated performance will supply 40 to 45% of the electrical load based on the performance of the first machines completed. Estimated payback time for the wind plant is 10 years at current electrical rates, but Valero has locked in the price of electricity from the wind plant. There are many other examples of distributed wind systems where the wind turbines are larger than 100 kW, but, because the power generated is used locally, they fall into the definition of distributed wind rather than as a wind farm, which sells all production at wholesale rates.

Figure 12.2 Wind turbines supplying electric power for the Valero Energy Corporation gasoline refinery. Example of large turbines installed as distributed wind systems.

COMMUNITY WIND

As stated earlier, community wind is a wind generation project owned collectively by a group of farmers, businesses, schools, or investors in a town or village. In many cases, the power is consumed near the place of generation and the investors share in the production or cost benefits. The shared production can either be electricity credit for electricity generated or by power sold locally. The majority of ownership benefits stay in the local community rather than be taken to some corporate office across the country and their stockholders. A key issue of community wind projects is the purchase agreement for the power generated. While many of these projects are used to offset energy purchases by the members of the investor group, other projects find equity partners who are eligible to receive the federal production tax credit, a tax credit that is paid for wind–generated electricity sold to an unrelated third party. It is important to benefit from net metering and the best purchase price possible. Access to low interest money is necessary for a successful project. Some of the larger projects use a Partnership Flip financial structure. A key component to this structure is that an outside investor who can utilize the income tax benefits becomes a "tax investor" and partner in the limited liability company and provides the majority of the capital. Majority ownership changes sometime after about 10 years from tax investor to local ownership when the tax benefits lapse or when the tax investor achieves his target rate of return. This structure reduces the capital that has to be raised by the local investors and allows for projects to be completed with local ownership. The wind education and advocacy group "Windustry" has developed a "community wind toolbox" that provides information and guidelines for establishing community wind projects [8].

One of the leading developers of community wind projects is Juhl Wind, Inc. located in Minnesota. They have completed about 20 community projects, with several of them using the flip financial structure [9]. At the beginning of 2005, most of the community wind projects were one or two turbines owned by a small group of local investors. However, over the years, the community wind projects have gotten larger, where it is not uncommon for them to involve 10 to 15 wind turbines with a capacity reaching 30 to 40 MW. This also means that the projects are moving from strictly a retail market to a wholesale market. In many ways they are becoming more like the typical wind farm operation, except for the ownership. Other developers are just getting started in developing these types of projects in other states where policies are being modified to allow the community wind structure.

SMALL WIND MACHINES

Small wind involves the most manufacturers and people using wind turbines of any of three segments of distributed wind systems. According to the 2012 Small Wind World Report conducted by the World Wind Energy Association [4], it is estimated that there are over 330 manufacturers of small wind machines in 40 countries. Canada, China, Germany, the United Kingdom, and the United States are the five leading manufacturers of small machines, accounting for over 50% of world production. The 2012 Small Wind World Report [4] reported that 656,084 small wind units were installed worldwide at the end of 2010 with 450,000 units or 60% of those in China (Figure 12.3). The United States was second with 144,000 units or 22%, and the United Kingdom was third with 21,610 or 3%. However, new markets in other parts of the world are opening up for small wind. 1.5 billion people still live without electricity in the world because most of them live in remote villages where it is prohibitively expensive to connect to the national electric grid [10]. It only takes a small amount of electricity to change the living conditions for many people in the world by providing lights, television,

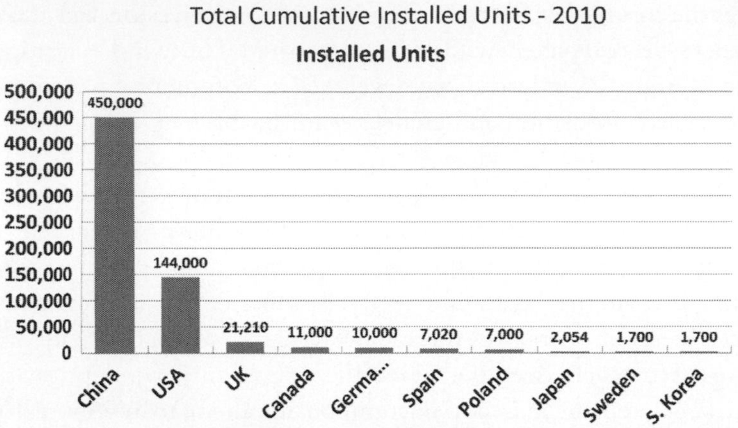

Total Cumulative Installed Units - 2010

Installed Units

Figure 12.3 Ninety-nine percent of the small wind machines were installed in these 10 countries at the end of 2010 [4].

and refrigeration for keeping food. Another impact on the sales of small machines is happening in Europe where several countries are changing policies allowing for increased prices for self-generated electricity. The United Kingdom and Germany are the first to create these new policies, but are being followed by Spain, Portugal, and France. The United States exported more small wind turbines in 2011 than were installed in the United States. The small wind industry is growing at about 25% per year worldwide, so as the prices for grid electric power continue to increase, we will continue to see small wind increase.

Other factors that contribute to the expansion of small wind systems include improved reliability, which is driven by increasing adoption of certification programs, and requirements that designs meet performance and safety standards. North America, Japan, the United Kingdom, and most of Europe require turbines to meet standards and obtain certification before being eligible for financial assistance or installation in restricted zones. Other countries are considering similar standards. One area that still needs improvement is the measurement and estimation of the wind resource for many rural or remote areas. Wind associations are helping to educate potential owners in resource measurement, siting, and operation. The American Wind Energy Association, in partnership with Wind Powering America, has instituted a program entitled "Wind for Schools," the focus of which is not the production of energy but the creation of a tool that can be utilized in the classroom to help educate students in science, technology, engineering, and mathematics (STEM). The program also gives local public power entities an opportunity to experience small wind firsthand, sometimes through installation help and interconnection support.

Eleven states actively participate in The Wind for Schools project, with approximately 100 turbines installed on school grounds [5]. Assistance comes from the individual states' land–grant university and typically an engineering professor, in this way offering college students an opportunity to learn more about wind technology and to participate in a small wind turbine installation. The Wind for Schools model includes a state facilitator who identifies specific K–12 schools that may be interested in hosting a small turbine. See www.windpoweringamerica.gov for more information about Wind for School projects.

In 2008, Jerome Middle School in Jerome, Idaho, installed a Skystream 3.7 wind turbine as part of its Wind for School project (Figure 12.4). The total cost of the installation, including the turbine and labor, was $16,275.50. To offset costs, the school sought in–kind donations for various aspects of the project, as well as grants to cover purchase of the turbine. Table 12.2 breaks down the project's donations, grants, and actual costs.

Figure 12.4 Skystream 3.7 wind turbine being installed at Jerome Middle School, Jerome, Idaho (NREL).

Table 12.2 Cost breakdown for Jerome Middle School Skystream 3.7 wind turbine

Item or task	Donor	Cost ($)
Permits	City of Jerome	272
Foundation	Starr Corporation	4,443
Tower	Jerome School District	3,450
Crane time	H&H Utility	3,000
Electrical wire	Portneuf Electrical	1,500
Electrician Labor	Power by Jake—Jake Cutler	360
Wind Turbine	Grant from Tidwell Foundation	3,250
Total Cost of Installation		16,275

As part of the Wind for Schools projects, school curricula were developed for grades K–12 and teacher training was provided. As well, in 2011 over 60 university students graduated with active participation in wind energy application centers. Education for children continues to play an important role in developing future markets and in the expansion of small wind systems.

REFERENCES

[1] Distributed Wind Energy Association web site: www.distributedwind.org July 20, 2012.
[2] International Electrotechnical Commission, Small Wind Standard, Design requirements for small wind turbines. IEC-61400-2, 2006-3.
[3] Web site for Juhl Wind, Inc. www.juhlwind.com, July 23, 2012.
[4] World Wind Energy Association. 2012 Small Wind World Report, March 2012.
[5] American Wind Energy Association. 2011 U.S. Small Wind Turbine Market Report, June 2012.
[6] Robson, N. Spirit Lake, Iowa, School District Honored for Wind Turbines. Sioux City Journal.com; June 20, 2012.
[7] Campoy, A. Valero Harnesses Wind Energy to Fuel Its Oil-Refining Process. WSJ.com; June 29, 2009.
[8] Community Wind at www.Windustry.org/community-wind/toolbox.
[9] Juhl Wind, Inc. www.juhlwind.com/projects.
[10] Rueter, G. Wind industry sees big potential for little turbines. Deutsche Welle, www.dw-world.de; April 19, 2011.

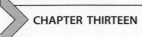

CHAPTER THIRTEEN

The Future of Small Wind
Factors That May or May Not Affect the Growth of Small Wind

Over the past ten years, small wind has seen steady growth, with the number of machines sold increasing yearly. However, it has not seen the rapid growth experienced by the large machine sector. Small wind industry leaders agree that several factors will influence the future of small wind, but they cannot agree on their relative importance. Siting and zoning policies may be more important in urban or more densely populated areas, whereas pricing and incentives may be more important in rural and remote areas. Also, the impact of these factors varies from country to country because of different policies and economics. The order in which these factors are discussed in this chapter is not meant to imply their relative importance.

ACCEPTANCE AND EDUCATION

As more and more wind systems are installed, the public is becoming more aware of the potential of wind as a supplement to the electrical generation capacity of local utility grids. For example, large wind farms have demonstrated a cost savings and economic boost to many local communities and regions. This favorable reaction is carrying over to small wind systems as well. In addition to more machines being visible to the general public, many schools have instituted educational programs in their science classes. The Wind Powering American program provides a K–12 curriculum on the principles of wind power [1]. Another popular program is the KidWind Project, which provides lessons and activities for middle schools. It includes kits for building systems and for learning how they operate. Several national wind energy associations also provide curricula for students in their respective countries. See "Wind Energy Curricula and Teaching Materials" on the Wind Powering America web site, www.windpoweringamerica.gov.

Another educational component is the training of technicians who will become the installers and maintenance workforce that sustains the small wind industry. Several vocational schools and universities have initiated training programs for wind technicians; however, most of them are geared toward the operation and maintenance of large utility-type machines. The small wind industry needs to engage these programs to include courses focused on small wind machines and their installation. The shortage of well-trained service personnel is and will continue to be a factor that may limit the growth of the small wind industry. Even with moderate growth, there could be a high demand for well trained technicians in the next 5 to 10 years.

Proper installation remains a major stumbling block facing successful operating systems. Wind turbines often perform exactly as designed and constructed, but because they are placed in poor locations, the actual energy produced is one-half to one-third of that anticipated. Well-trained and highly qualified installers are critical to the success of any small wind project. The industry needs to strengthen the installer training and certification program through organizations such as the North American Board of Certified Energy Practitioners [2] or similar organizations worldwide.

NEW OR REVISED POLICIES AND INCENTIVES

The impact of energy policies has played a significant role in the growth or slowdown of renewable energy technologies over the years. This is especially true for wind energy: when some early incentives based on rated capacity were in place, the industry experienced rapid growth. When those incentives and policies were changed, the wind industry almost died for a few years until new policies were initiated. What is needed are longer-term incentives and policies that are designed for gradual transition into new policies. Abrupt changes cause rapid expansion in manufacturing to meet demand but then create slowdowns and layoffs because of a lack of demand for new wind machines. Many countries and states have policies in place that mandate electric utilities to generate a certain percentage of their electric power from wind or a combination of renewable energy sources by some projected date. These policies have helped to create interest in wind systems, but it is incentives that help to increase sales of wind systems.

Incentives that reduce the price of wind machines stimulate sales and thus growth in the small wind industry. In the United States, a 30% tax credit on the purchase of small wind turbines has provided for some growth in the industry, but it has not been as effective as state programs that provide cash rebates or grants that provide direct cash toward the purchase of the wind machine. Tax credits benefit only those who pay high taxes and generally do not encourage or provide much incentive to the average person who just wants to cut electricity costs and create a hedge against future price increases. Again, short one- or two-year programs do not really support the growth needed to stimulate the industry. Nor does a single incentive appear to reach all potential wind energy enthusiasts; therefore, a variety of incentives may be needed to broaden the group of potential wind turbine buyers.

Renewable energy credits (RECs) represent another incentive that has not been fully utilized by small wind turbine owners. RECs are available when producing grid-connected electricity, but they are not used because their value is not worth the effort or the paper the contracts are written on. The value of the REC from a 1 kW small wind turbine is one-thousandth of the value of that for a 1 MW wind machine. If it were possible to group several small machines, a large enough package might become valuable enough to be bought and sold. However, the RECs may remain for larger developments and wind farm activities.

Another policy change that could positively impact small wind is a broader access to utility interconnections. Interconnection policies and costs vary greatly from utility to utility because different regulations are imposed on investor-owned utilities as compared to cooperatives. Publicly owned utilities are usually not subject to the regulations that govern other utilities. Consider as an example a school system in a rural area that installs wind turbines at three campuses. Two of the campuses develop interconnection agreements with an investor-owned utility that supplies electricity to the main area of town, which provides a good return for excess electricity generated. The third campus interconnection agreement is with the local rural electric cooperative, but its turbine has not proven economical because of limited power production and lower payback rates. This turbine is turned off several days per month because it has already reached its limit on generating excess power back into the grid. Simpler and uniform interconnection policies can have a large influence on the future of small wind systems and the economics of small wind.

CHANGING ECONOMICS

That the cost of electricity will continue to increase is almost an understood reality; however, the rate of increase is difficult to predict. Rates are localized, and some locations have doubled in the last 20 years while others in the same period have gone up only slightly. Still, all have increased some because they are based on the cost of fuel to operate the generating plants. The demand for electricity is increasing at a rapid pace, which requires more generating capacity, which leads to higher electricity prices. Almost everyone agrees that electricity prices will continue to increase at a rapid rate. At some point during the lifetime of a wind turbine, the electricity rate could easily double, which means that the owner could save more than 50% on his or her electric bill each month. It is on these uncertain changes that the wind turbine owner is risking his or her capital investment. Higher electric rates are an incentive to invest in wind energy and fix the owner's cost of electricity at the cost of the wind turbine installation.

The cost of purchasing a wind turbine will probably continue to increase over time because of increasing costs of system components. There should be some cost saving in the production of wind turbines when a point of producing more systems is reached. Manufacturers should be able to purchase components in bulk, thus saving some cost per item and in manufacturing. All of these factors will influence the economics of owning and operating small wind systems.

IMPROVED AND ADVANCED TECHNOLOGY

There have been significant improvements in small turbine technology during the last 8 to 10 years; however, there is still plenty of opportunity for more. One main improvement has been in the use of the permanent magnet alternator or generator in place of DC generators for obtaining a DC output from smaller wind systems. There is a large gain in efficiency with the permanent magnet alternator and the system speed control is much easier, thus reducing periods of non-production. As grid-connected wind turbines increase in size from 1 to 2 kW and approach 5 to 7 kW, the choice to use an induction generator may be made because of total cost. This creates a new design model that may or may not be as efficient as current models. Most certainly, it will change the current trend in that most small machines use permanent magnet alternators and inverters. So far, not many manufacturers are interested in entering the 25 to 50 kW range of small wind machines. This is simply because of the fairly steady supply of refurbished machines coming from repowered wind farms in California and Denmark. This supply of used machines will soon be exhausted, and the need for new machines may create a place for some new development in this size range.

Rotor blades are one area that has been mainly overlooked for new technology development. Much work has been done on blade materials, but little attention has been applied to airfoil shape and performance efficiency. One reason that only a few manufacturers have put much effort into airfoil shape is the development cost involved in building test blades and conducting tests. Controlled testing in wind tunnels is expensive and requires follow-up testing at several wind speeds to verify results. Acoustic measurements are also necessary to meet the new performance standards that are in place.

Advances in wireless data communication via computers, smart phones, and the Internet have provided new opportunities for better and more sophisticated controls for small wind systems. The use of wireless technology via the Internet allows for real-time data monitoring of performance and operation. It also allows for remote-controlled starting and stopping of the wind turbine to avoid adverse weather conditions. These same controller systems may incorporate self-trouble shooting software that can relay information to service personnel so that trips to remote sites are minimized and repairs are completed rapidly, thus reducing downtime.

STANDARDS AND CERTIFICATION FOR TURBINES

The first international safety standards for small wind turbines were published in 1996, but many manufacturers paid little attention to them because of the detailed specialized analyses and testing they required. It was not until these standards were revised in 2006, with simplified design equations based on recent testing and research, that they have been utilized by manufacturers [3]. Also, several organizations and country wind associations have developed certification programs based on the 2006 international design standards. Most certification programs were implemented after 2010; therefore, the concept and processes of certification are just now becoming the accepted methods of operation. The main focus of the standards and certification is to improve the reliability and performance of small wind turbines. Several wind turbines have been certified, and many others are at some stage of applying and testing under the various certification bodies.

A committee of the International Electrotechnical Commission (IEC) is developing a revised standard for small wind turbines, and the initial draft is out for reviews and comments. Both American and British wind energy associations are developing revisions to their standards to make them compatible with the new IEC standard. The IEC standard focuses on design and performance; individual country standards deal with safety, performance, and acoustics. One important aspect of revised standards is that they are becoming very similar and that once a wind turbine is certified to a specific standard, it can be easily certified to most of the others. This will bring uniformity to the small wind turbine industry and ensure that wind turbines operate safely and reliably. A uniform certification process, once in place, will open the whole world to any manufacturer who has a certified wind turbine. Certification certainly changes the dynamics of marketing and selling wind turbines and it should broaden the opportunities for small manufacturers.

Some industry long-timers think that certification may bring a consolidation of wind turbine sizes, but the author supports the idea that the market will determine sizes. We already see more machines in the 4 to 5 kW range because of need for a larger unit connected to the electric grid that will meet almost 100 % of typical home usage. The 1 to 2 kW machines will remain strong for the off-grid market because of ease of installation and transportation of components. As for machines larger·than 10 kW, the author does not see a large number entering the market and those that do will be based on either a market size or a swept-rotor size, as they were during the 1980s [4]. Many were based on the 7.5 m blade developed by Danish company Aerostar, and others chose a 10 m rotor diameter.

SUPPLY CHAIN CHANGES

One area that will change the most and may greatly influence the growth of small wind turbines and wind systems is the supply chain. There are two components to the supply chain, and each can potentially influence the growth and probably the rate of growth as much as any other factor. One aspect of the supply chain is the components required to produce a small wind system. More and more, manufacturers are using off-the-shelf components rather than those manufactured in-house because of quality control and cost of production. Therefore, it is important to have contracts in place to ensure that enough products are being shipped in a timely manner for assembly into the finished wind turbine. This will become more critical as production increases and as manufacturers go from producing one or two machines a month to one or two machines per week and even per day.

The other aspect of the supply chain is the marketing of turbines. Today, the majority of machines are sold directly from the manufacturer to a customer or a specialized installer. In the future, turbines may be sold through a dealer network similar to those for cars, or, much like household appliances, in large box stores. Dealers would offer an installation package through a contract with a certified installer. Thus the new supply chain would flow from the manufacturer to a seller, with or without an installation package. As we see these changes take place in the supply chain, it is hoped that the quality of the product and the quality of the installation will improve to provide the best product available.

GLOBAL MARKETING

The United States has been the leader in installed capacity of small wind systems, but China has been the leader in the number of small wind systems sold. The reason for U.S. leadership in capacity is the number of machines installed in 25 to 65 kW sizes, whereas China has been mostly installing machines less than 1 kW. In the last year, the market has seen its greatest growth in countries like the United Kingdom, Italy, Japan, and Taiwan. In fact, all of Europe is seeing a growth in the installation of small wind systems that are grid intertied using wind systems in the 2 to 5 kW range. Again, the size of the wind turbine sold and installed in a particular country depends greatly on whether the machine will be operated as grid-tied or a stand-alone system. Most grid-tied systems for individual homes will be in the 2 to 5 kW size; stand-alone systems will be 0.5 to 1 kW. Turbines that are 10 kW and larger will be mainly used for commercial applications including farms and ranches.

If there is a significant breakthrough in the cost of systems, many less-developed countries will begin to develop more wind power. Recently, several countries have greatly expanded their development of wind systems for large electric power generation. This growth will expand to include small wind as well.

REFERENCES

[1] Wind Powering America, www.windpoweringamerica.gov.
[2] NABCEP, North American Board of Certified Energy Practitioners, www.nabcep.org.
[3] International Standard. Design Requirements for Small Wind Turbines. IEC61400-2. Geneva, Switzerland: International Electrotechnical Commission; 2006.
[4] Spera David A. Wind Turbine Technology. New York: ASME Press; 1994.

Useful Web Sites

NATIONAL WIND ENERGY ASSOCIATIONS

American Wind Energy Association, www.awea.org
Canadian Wind Energy Association, www.canwea.ca
RenewableUK, (formerly British Wind Energy Association),
    www.renewableUK.com
The Windpower, www.thewindpower.net (lists wind energy associations)
World Wind Energy Association, www.wwindea.org
Distributed Wind Energy Association, www.distributedwind.org

INTEREST AREAS

Community Wind, www.Windustry.org/community-wind
Database of State Incentives for Renewables and Efficiency (DSIRE),
    http://dsireusa.org
Midwest Renewable Energy Association, www.midwestrenew.org
North American Board of Certified Energy Practitioners (NABCEP),
    www.nabcep.org
Use the Wind, www.Usethewind.net (wind photographs)

GOVERNMENT

Wind Powering America, www.windpoweringamerica.gov
National Renewable Energy Laboratory, National Wind Technology
    Center, www.nrel.gov/wind/nwtc
Federal Energy Regulatory Commission (FERC), www.ferc.gov/
    industries/electric
National Grid, www.nationalgridus.com/niagaramohawk

CERTIFICATION AND STANDARDS

Small Wind Certification Council, www.smallwindcertification.org

U.K. Microgeneration Certification Scheme, www.microgeneration certification.org

Germanischer Lloyd Industrial, GL Renewables Certification, www.gl-group. com/en/certification/renewables

IEEE Standard 1547, *Interconnecting Distributed Resources with Electric Power Systems*, www.ieee.org

National Fire Protection Association, National Electric Code (2011 ed., NFPA 70), www.nfpa.org

UL Standard 1741, *Inverters, Converters, Controllers and Interconnection System Equipment for Use with Distributed Energy Resources*, http://ulstandardsinfonet. ul.com

OTHERS

Alternative Energy Institute, West Texas A&M University, www.windenergy.org

Wind Energy Institute of Canada, WEICan, www.weican.ca

Intertek Wind Turbine Testing, www.intertek.com/wind

INDEX

Note: Page numbers with "f" denote figures; "t" tables; "b" boxes.